HOUGHTON MIFFLIN HARCOURT

Math Expressions Common Core

Dr. Karen C. Fuson

GRADE 4

Volume 2

This material is based upon work supported by the
National Science Foundation
under Grant Numbers
ESI-9816320, REC-9806020, and RED-935373.

Any opinions, findings, and conclusions, or recommendations expressed in this material are those of the author and do not necessarily reflect the views of the National Science Foundation.

Cover Credit: (Polar bear) ©imagebroker.net/SuperStock

Printed in the U.S.A.

ISBN: 978-0-547-82441-3

8 9 10 0868 21 20 19 18 17 16 15 14

4500478331 A B C D E F G

VOLUME 2 CONTENTS

UNIT 5 Measurement

BIG IDEA 1 Converting Measurements

BIG IDEA 2 Perimeter and Area

UNIT 6 Fraction Concepts and Operations

BIG IDEA 1 Fractions with Like Denominators

UNIT 7 Fractions and Decimals

BIG IDEA 3 Analyzing Quadrilaterals

BIG IDEA 4 Analyzing Polygons

STUDENT RESOURCES

Dear Family,

This unit is about the metric measurement system. During this unit, students will become familiar with metric units of length, capacity, mass, and time, as well as the size of each when compared to each other.

Share with your family the Family Letter on Activity Workbook page 47.

One **meter** is about the distance an adult man can reach, or a little longer than a yard.

One **liter** is about two large glasses of liquid, or a little more than a quart.

One **gram** is about the mass of a paper clip or a single peanut. One **kilogram** is a little more than 2 pounds.

Students will also discover that the metric system is based on multiples of 10. Prefixes in the names of metric measurements tell the size of a measure compared to the size of the base unit.

Units of Length

kilometer	hectometer	decameter	meter	decimeter	centimeter	millimeter
km	hm	dam	m	dm	cm	mm
10 × 10 × 10 × larger	10 × 10 × larger	10 × larger	1 m	10 × smaller	10 × 10 × smaller	10 × 10 × 10 × smaller
1 km = 1,000 m	1 hm = 100 m	1 dam = 10 m		10 dm = 1 m	100 cm = 1 m	1,000 mm = 1 m

The most commonly used length units are the **kilometer, meter, centimeter**, and **millimeter**.

The most commonly used capacity units are the **liter** and **milliliter**.

The most commonly used units of mass are the **gram, kilogram**, and **milligram**.

If you have any questions or comments, please call or write to me.

Sincerely,
Your child's teacher

This unit includes the Common Core Standards for Mathematical Content for Measurement and Data, 4.MD.1, 4.MD.2, 4.MD.3, 4.MD.4 and all Mathematical Practices.

Estimada familia:

Esta unidad trata del sistema métrico de medidas. Durante esta unidad, los estudiantes se familiarizarán con unidades métricas de longitud, capacidad y masa, así como con el tamaño de cada una comparada con las otras.

Un **metro** es aproximadamente la distancia que un hombre adulto puede alcanzar extendiendo el brazo, o un poco más de una yarda.

Un **litro** es aproximadamente dos vasos grandes de líquido, o un poco más de un cuarto de galón.

Muestra a tu familia la Carta a la familia de la página 48 del Cuaderno de actividades y trabajo.

Un **gramo** es aproximadamente la masa de un clip o un cacahuate. Un **kilogramo** es un poco más de 2 libras.

Los estudiantes también descubrirán que el sistema métrico está basado en múltiplos de 10. Los prefijos de los nombres de las medidas métricas indican el tamaño de la medida comparado con el tamaño de la unidad base.

Unidades de longitud

kilómetro	hectómetro	decámetro	metro	decímetro	centímetro	milímetro
km	hm	dam	m	dm	cm	mm
10 × 10 × 10 × más grande	10 × 10 × más grande	10 × más grande	1 m	10 × más pequeño	10 × 10 × más pequeño	10 × 10 × 10 × más pequeño
1 km = 1,000 m	1 hm = 100 m	1 dam = 10 m		10 dm = 1 m	100 cm = 1 m	1,000 mm = 1 m

Las unidades de longitud más comunes son **kilómetro, metro, centímetro** y **milímetro.**

Las unidades de capacidad más comunes son **litro** y **mililitro.**

Las unidades de masa más comunes son **gramo, kilogramo** y **miligramo.**

Si tiene alguna pregunta o algún comentario, por favor comuníquese conmigo.

Atentamente,
El maestro de su niño

Esta unidad incluye los Common Core Standards for Mathematical Content for Measurement and Data, 4.MD.1, 4.MD.2, 4.MD.3, 4.MD.4 and all Mathematical Practices.

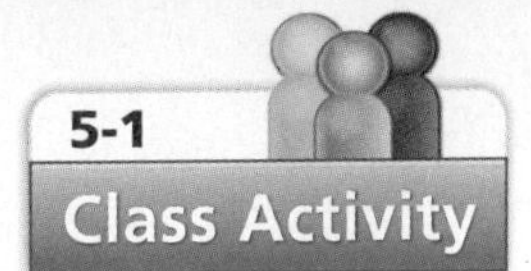

VOCABULARY
millimeter
centimeter
decimeter
meter

► Parts of a Meter

Find these units on your meter strip.

mm	10	20	30	40	50	60	70	80	90	100
cm	1	2	3	4	5	6	7	8	9	10
dm										1
m										

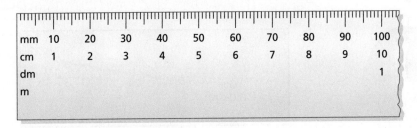

1. Find one **millimeter** (1 mm) on your strip.
   What objects are about 1 mm wide?

2. Find one **centimeter** (1 cm) on your strip.
   How many millimeters are in 1 cm?

3. What objects are about 1 cm wide?

4. Find one **decimeter** (1 dm) on your strip.
   How many centimeters are in 1 dm?

   This is one **meter** (1 m) that has been folded into decimeters to fit on the page.

5. How many decimeters are in 1 m?

► Choose Appropriate Units

Record which metric unit of length is best for measuring each object. Be prepared to justify your thinking in class.

6. 

7. 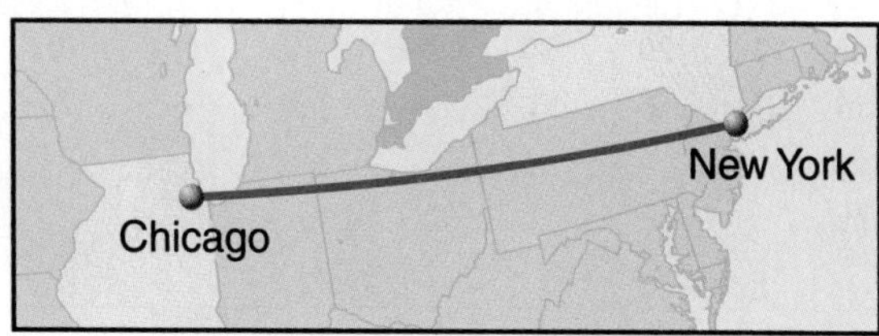

8. 

9. 

10. 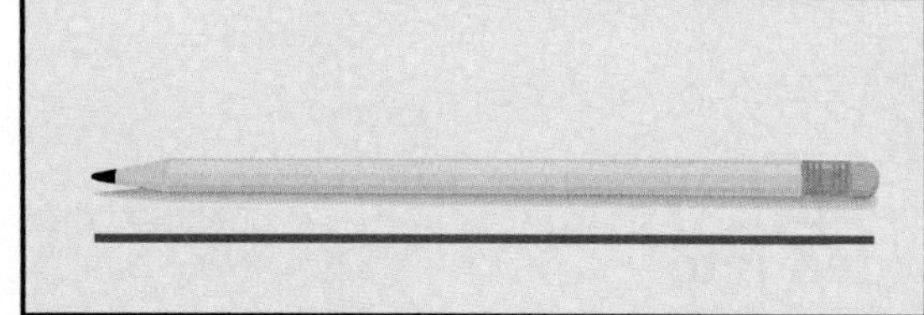

11. 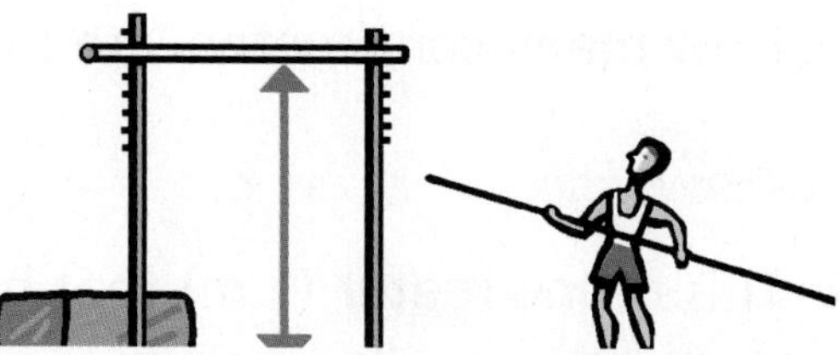

12. 

13. 

VOCABULARY
kilometer
prefixes
metric system

► Metric Prefixes

Units of Length

kilometer	hectometer	decameter	meter	decimeter	centimeter	millimeter
km	hm	dam	m	dm	cm	mm
10 × 10 × 10 × larger	10 × 10 × larger	10 × larger	1 m	10 × smaller	10 × 10 × smaller	10 × 10 × 10 × smaller
1 km = 1,000 m	1 hm = 100 m	1 dam = 10 m		10 dm = 1 m	100 cm = 1 m	1,000 mm = 1 m

14. What words do you know that can help you remember what the **prefixes** mean in the **metric system**?

15. How do the lengths of the different units relate to each other?

16. How many meters are in 1 **kilometer**?

17. How many millimeters are in 1 m?

18. How many centimeters are in 1 m?

19. What makes the metric system easy to understand?

Use Activity Workbook page 49.

▶ Convert Metric Units of Measure

You can use a table to convert measurements.

20. How many decimeters are in one meter? ______

21. Complete the equation.
1 meter = ______ decimeters

22. Complete the table. Explain how you found the number of decimeters in 8 meters.

Meters	Decimeters
2	2 × 10 = 20
4	___ × 10 = ___
6	6 × ___ = ___
8	___ = ___

You can also use a number line to convert measurements.

23. Complete the equation. 1 kilometer = ______ meters

24. Label the double number line to show how kilometers (km) and meters (m) are related.

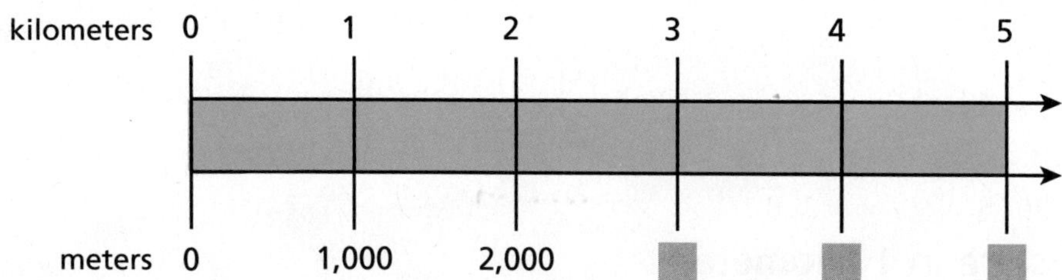

Solve each problem. Label your answers with the correct units.

25. Marsha drove her car 6,835 kilometers last year. How many meters did Marsha drive last year?

26. John's television is 160 cm wide. How many millimeters wide is the television?

Solve.

27. 5 m = ______ cm

28. 3 hm = ______ m

29. 7 km = ______ m

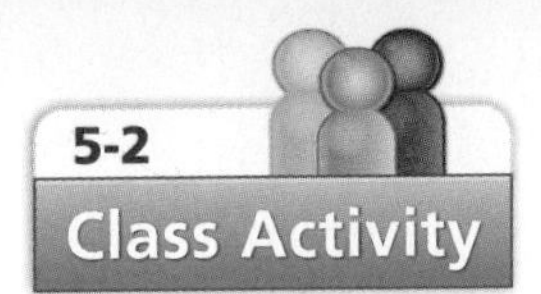

Use Activity Workbook page 50.

VOCABULARY
liquid volume
liter
milliliter
kiloliter

▶ Measure Liquid Volume

The base metric unit of **liquid volume** is a **liter**.

Units of Liquid Volume

kiloliter	hectoliter	decaliter	liter	deciliter	centiliter	milliliter
kL	hL	daL	L	dL	cL	mL
10 × 10 × 10 × larger	10 × 10 × larger	10 × larger	1 L	10 × smaller	10 × 10 × smaller	10 × 10 × 10 × smaller
1 kL = 1,000 L	1 hL = 100 L	1 daL = 10 L		10 dL = 1 L	100 cL = 1 L	1,000 mL = 1 L

Ms. Lee's class cut a two-liter plastic bottle in half to make a one-liter jar. They marked the outside to show equal parts.

1. How many **milliliters** of water will fit in the jar?

2. How many of these jars will fill a **kiloliter** container? Explain why.

You can use a table or a double number line to convert units of liquid measure.

3. Complete the table.

Liters	Deciliters
3	3 × 10 = 30
5	▢ × 10 = ▢
7	7 × ▢ = ▢
12	▢ = ▢

4. Label the double number line to show how liters (L) and milliliters (mL) are related.

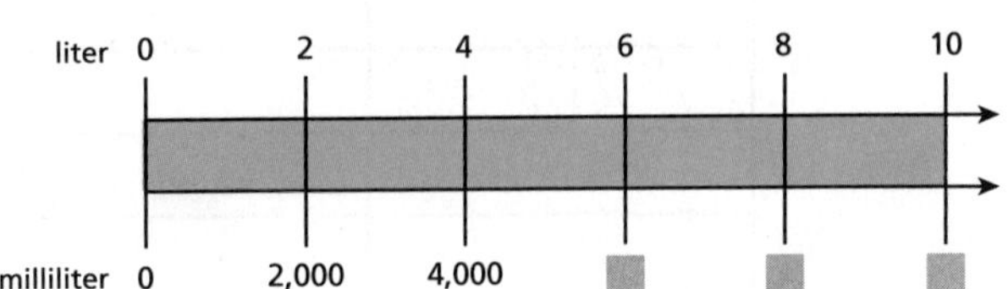

Show your work on your paper or in your journal.

▶ What's the Error?

Dear Math Students,

Today I had to solve this problem.

Meredith wanted to make some punch for a party. The recipe to make the punch called for 3 liters of fruit juice, 2 liters of apple juice, and 1 liter of grape juice. How many milliliters of juice is needed for the recipe? I said that the recipe calls for 600 milliliters of juice. Here is how I solved the problem.

3 L + 2 L + 1 L = 6 L x 100 = 600 mL

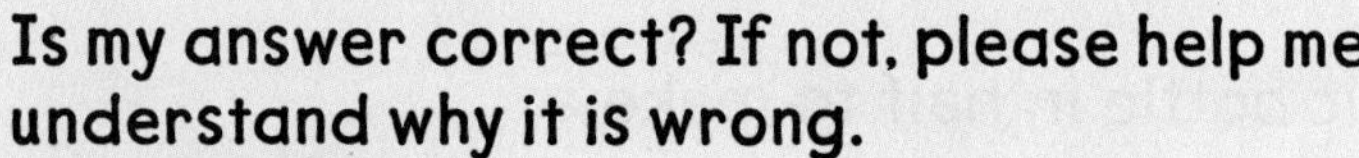

Is my answer correct? If not, please help me understand why it is wrong.

Your friend,
Puzzled Penguin

5. Is the Puzzled Penguin correct? Explain your thinking.

6. Draw a table to show the conversion from liters to milliliters for each type of juice needed for the recipe.

Liters	Milliliters

7. Name another way that you could show the conversion from liters to milliliters for each type of juice.

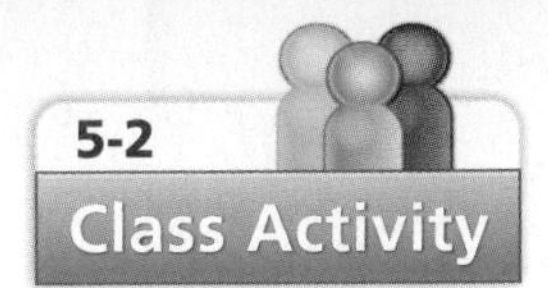

VOCABULARY
mass
gram
kilogram
milligram

► Measure Mass

The basic unit of **mass** is the **gram**.

Units of Mass

kilogram	hectogram	decagram	gram	decigram	centigram	milligram
kg	hg	dag	g	dg	cg	mg
10 × 10 × 10 × larger	10 × 10 × larger	10 × larger	1 g	10 × smaller	10 × 10 × smaller	10 × 10 × 10 × smaller
1 kg = 1,000 g	1 hg = 100 g	1 dag = 10 g		10 dg = 1 g	100 cg = 1 g	1,000 mg = 1 g

8. How many **milligrams** are equal to 1 gram?

9. How many grams are equal to 1 kilogram?

If you weighed 1 mL of water, you would find that its mass would be one gram (1 g).

10. Is the gram a small or large unit of measurement? Explain your thinking.

You can use a table or a double number line to convert units of mass.

11. Complete the table.

Grams	Centigrams
4	4 × 10 = 40
8	▢ × 10 = ▢
12	12 × ▢ = ▢
15	▢ = ▢

12. Label the double number line to show how kilograms (kg) and grams (g) are related.

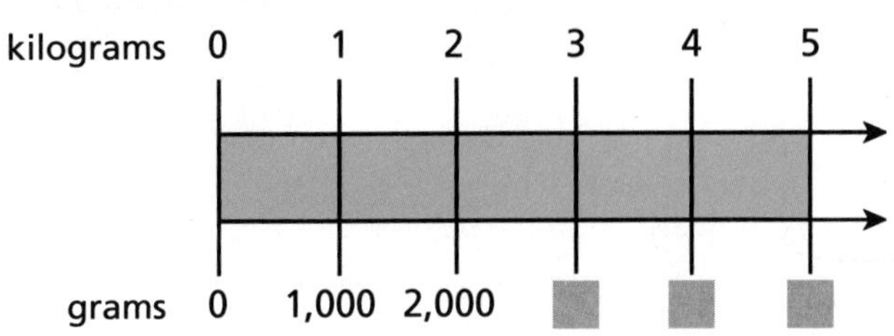

► Practice Converting Metric Units

Solve.

13. Martin measured the mass in grams of four different objects and recorded the information in the table below. Complete the table to find the mass of each object in milligrams.

Grams	Milligrams
4	4,000
7	
11	
15	

14. Olivia bought four different-sized containers and filled them each with water. She recorded the liquid volume of each container in liters below. Complete the table to find the liquid volume of each container in centiliters.

Liters	Centiliters
1	
3	
4	400
6	

15. Hayden has a crayon with a mass of 8 grams. Complete the double number line to find the mass of the crayon in centigrams.

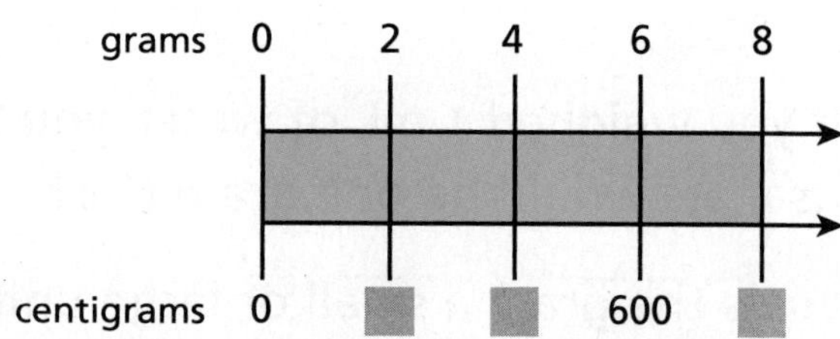

16. Jennifer buys a 2-liter bottle of apple juice and a 3-liter bottle of orange juice at the market. How many deciliters of juice does Jennifer buy in all?

17. Elena has a cat with a mass of 4 kilograms. Ginger's cat has a mass that is 2 times as much as Elena's cat. What is the mass of Ginger's cat in grams?

▶ Basic Units of Time

Complete the table.

Units of Time	
1. 1 minute = ▇ seconds	5. 1 year = ▇ days
2. 1 hour = ▇ minutes	6. 1 year = ▇ weeks
3. 1 day = ▇ hours	7. 1 year = ▇ months
4. 1 week = ▇ days	8. 1 leap year = ▇ days

▶ Convert Units of Time

Complete the table.

9.

Days	Hours
1	24
2	
3	
4	

10.

Hours	Minutes
1	60
3	
5	
7	

11.

Years	Months
3	36
6	
9	
12	

12.

Hours	Seconds
1	3,600
2	
3	
4	

Solve.

13. 36 min = ____ sec

14. 41 days = ____ hours

15. 72 hours = ____ min

16. 16 weeks = ____ days

17. 6 years = ____ days

18. 2 weeks = ____ hours

VOCABULARY
line plot

▶ Make a Line Plot

A **line plot** displays data above a number line. Jamal asked his classmates about the time they spend reading. He organized the answers in the table.

Time Spent Reading	Number of Students
0 hour	0
$\frac{1}{4}$ hour	2
$\frac{1}{2}$ hour	5
$\frac{3}{4}$ hour	4
1 hour	4

19. Use the table to complete the line plot.

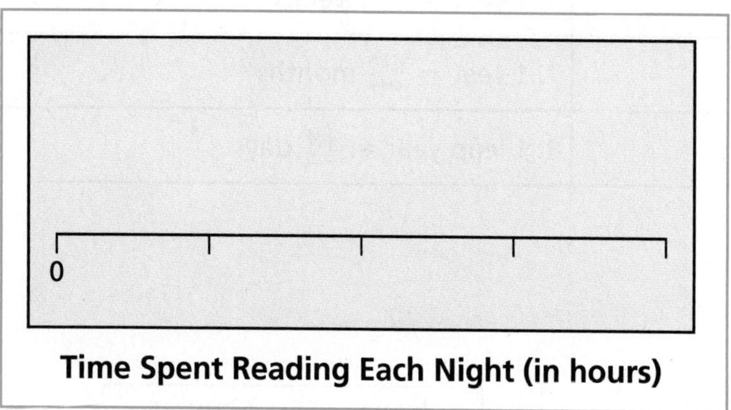

Time Spent Reading Each Night (in hours)

20. How many classmates did Jamal ask about time spent reading?

21. What amount of time had the most responses?

▶ Practice

Solve.

22. Fiona asked her friends how much time they spend using a computer at home each night. Use the information in the table to make a line plot.

Time Spent on Computer	Number of Students
0 hour	4
$\frac{1}{4}$ hour	4
$\frac{1}{2}$ hour	7
$\frac{3}{4}$ hour	3
1 hour	9

23. Marissa wants to know how many minutes she has practiced the piano. Label the double number line to show how hours and minutes are related. How many minutes has she practiced if she practiced for 4 hours?

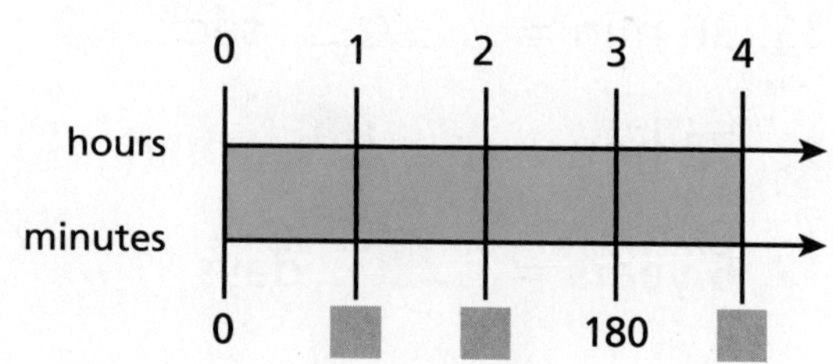

▶ Read Elapsed Time

You can imagine the hands moving to tell how much time has passed.

How many hours have passed since 12:00 on each clock?
How many minutes?

24. 

25. 

26. 

How many hours and how many minutes does the clock show? Write the time that the clock shows.

27. 

_____ hours

_____ minutes

28. 

_____ hours

_____ minutes

29. 

_____ hours

_____ minutes

How many hours and minutes have passed between the times shown in:

30. Exercises 27 and 28

31. Exercises 28 and 29

32. Exercises 27 and 29

Show your work on your paper or in your journal.

► Solve Elapsed-Time Problems

Solve.

33. The school store is open for 1 hour and 45 minutes in the afternoon. The store closes at 2:55 P.M. What time does the school store open?

34. Hannah's practice starts are 9:45 A.M. and ends at 12:35 P.M. How long is Hannah's practice?

35. Bella's dance class starts at 3:05 and lasts 1 hour 35 minutes. At what time does Bella's dance class end?

36. Lynn goes to the mall with her mom. They get to the mall at 6:20 and leave at 8:25. How long were Lynn and her mom at the mall?

37. It takes Reese 32 minutes to walk to school. She gets to school at 7:55 A.M. What time did Reese leave her house to walk to school?

38. Kevin is baking a cake. He starts making the cake at 8:03 P.M. It takes him 1 hour 17 minutes to finish making the cake. At what time was Kevin finished making the cake?

5-4
Class Activity

VOCABULARY
inch
foot
yard
mile

► Units of Length

1. This line segment is 1 **inch** long. Name an object that is about 1 inch long.

2. One **foot** is equal to 12 inches. Name an object that is about 1 foot long.

3. One **yard** is equal to 3 feet or 36 inches. Name an object that is about 1 yard long.

4. Longer distances are measured in miles. One **mile** is equal to 5,280 feet or 1,760 yards. Name a distance that is about 1 mile long.

► Convert Customary Units of Length

5. Complete the table.

Feet	Inches
1	12
2	
3	
4	
5	

6. Complete the table.

Yards	Feet
2	6
4	
6	
8	
10	

Solve.

7. 9 yards = _____ inches

8. 26 feet = _____ inches

9. 4 miles = _____ feet

10. 2 miles = _____ inches

▶ Measure Length

Write the measurement of each line segment to the nearest $\frac{1}{8}$ inch.

11.

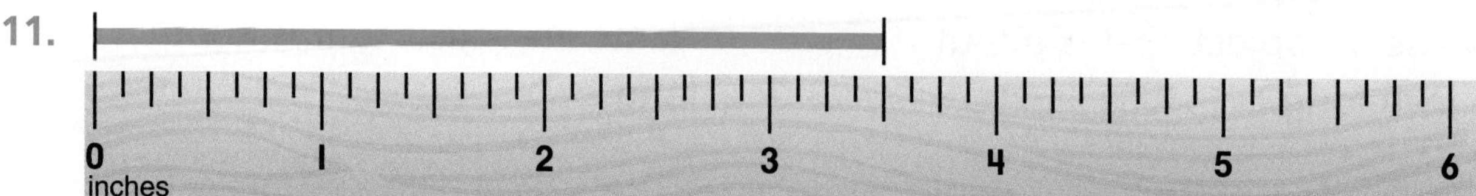

12.

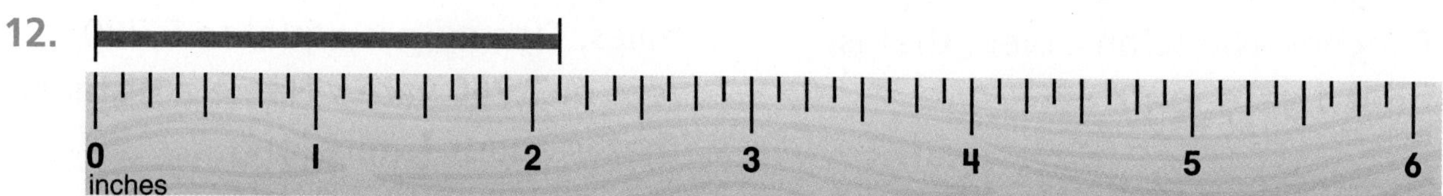

13.

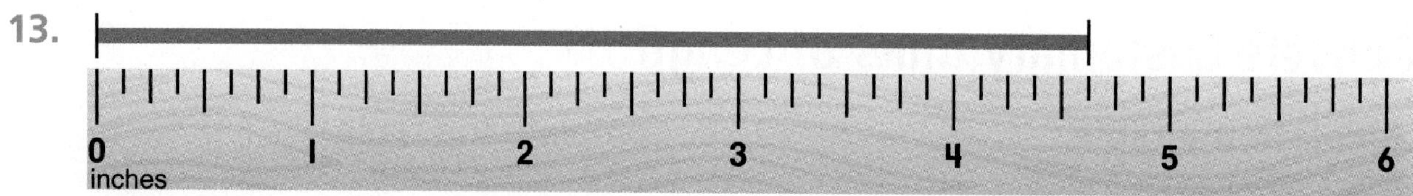

14.

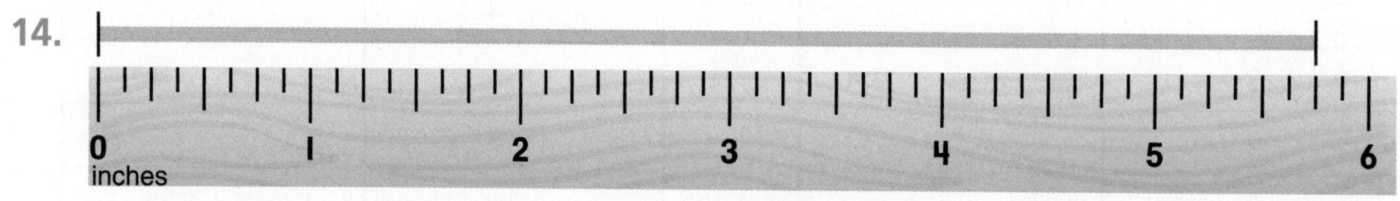

15.

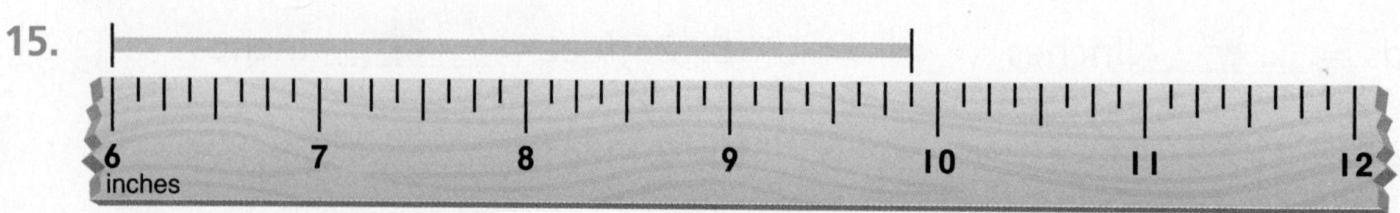

VOCABULARY
pound
ounce

▶ Pounds and Ounces

The pound is the primary unit of weight in our customary system. One **pound** is equal to 16 **ounces**.

Butter and margarine are sold in 1-pound packages that contain four separately wrapped sticks.

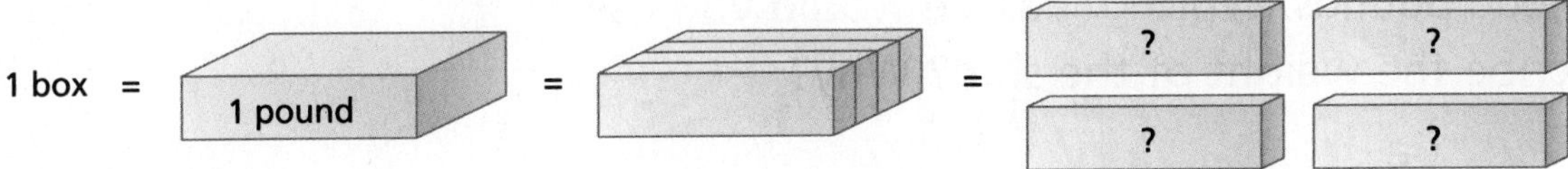

1. What is the weight in ounces of one box?

2. What is the weight in ounces of one stick?

3. Kimba buys a bag of flour that weighs 5 pounds. Complete the table. How many ounces are equal to 5 pounds?

Pounds	Ounces
1	16
2	
3	
4	
5	

4. Describe how to convert pounds to ounces without using a table.

5. When Martin weighed his dog in April, the dog weighed 384 ounces. When he weighed the dog in August, the dog weighed 432 ounces. How many ounces did Martin's dog gain between April and August?

5-5
Class Activity

Show your work on your paper or in your journal.

VOCABULARY
ton

▶ Tons

The weight of heavy items such as cars, trucks, boats, elephants, and whales is measured in **tons**. One ton is equal to 2,000 pounds.

6. A ship weighs 12,450 tons. In pounds, the ship weighs 24,900,000 pounds. Which measure would you use to describe the weight of the ship? Why?

7. A trailer can carry 2 tons of cargo. How many pounds of cargo can the trailer carry?

8. One cargo container weighs 4 ounces. A shipment of cargo containers weighs a total of 1 ton. How many cargo containers are in the shipment? Show your work.

▶ Practice

Solve.

9. 3 tons = ________ pounds

10. 7 pounds = ________ ounces

11. 5 tons = ________ pounds

12. 12 pounds = ________ ounces

13. 9 tons = ________ pounds

14. 19 pounds = ________ ounces

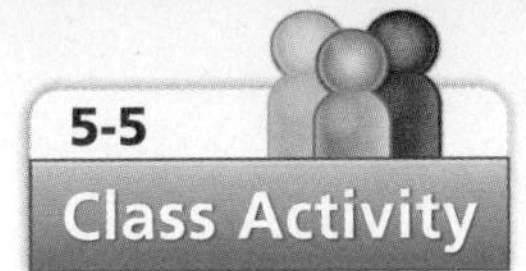

VOCABULARY
cup
fluid ounce
quart
pint
gallon

► Liquid Volume

In the customary system, the primary unit of liquid volume is a **cup**.

1 cup = 8 **fluid ounces**

4 cups = 1 **quart**

2 cups = 1 **pint**

4 quarts = 1 **gallon**

15. Complete the table.

Quarts	Fluid Ounces
1	32
2	
3	
4	
5	
6	

16. Label the double number line to show how gallons (gal) and cups (c) are related.

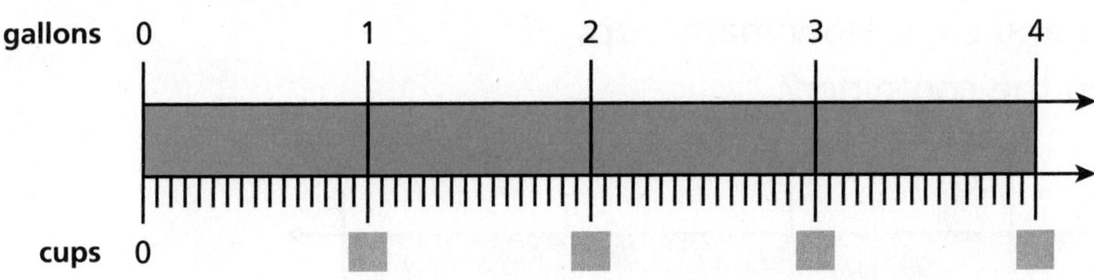

Solve.

17. 3 qt = ______ c

18. 10 c = ______ fl oz

19. 2 gal = ______ pt

► Solve Real World Problems

Solve.

20. A race is 5 miles long. Complete the table. How many feet are equal to 5 miles?

Miles	Feet
1	5,280
2	
3	
4	
5	

21. DeShawn has a watermelon that weighs 64 ounces. He cuts the watermelon into 8 pieces that each have the same weight. How many ounces does each piece of watermelon weigh?

22. A container has 4 quarts of water left in it. Complete the double number line to show the relationship between quarts and cups. How many cups of water are left in the container?

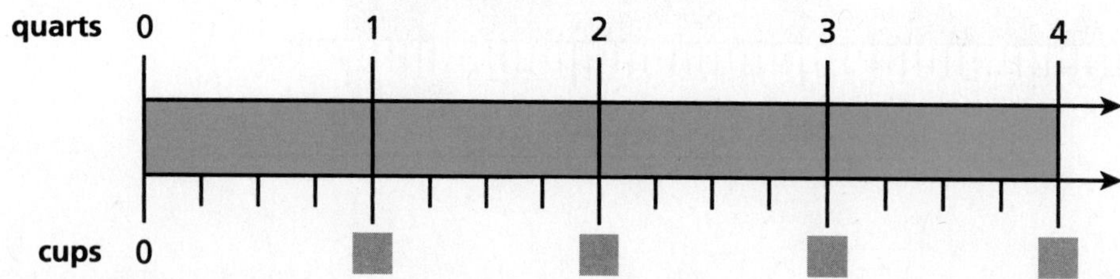

23. Melinda has 2 yards of fabric to make a banner. How many feet of fabric does she have? How many inches of fabric does she have?

VOCABULARY
perimeter
length
width
formula

▶ Units of Perimeter

The prefix *peri-* means "around." The suffix *-meter* means "measure." **Perimeter** is the measurement of the distance around the outside of a figure.

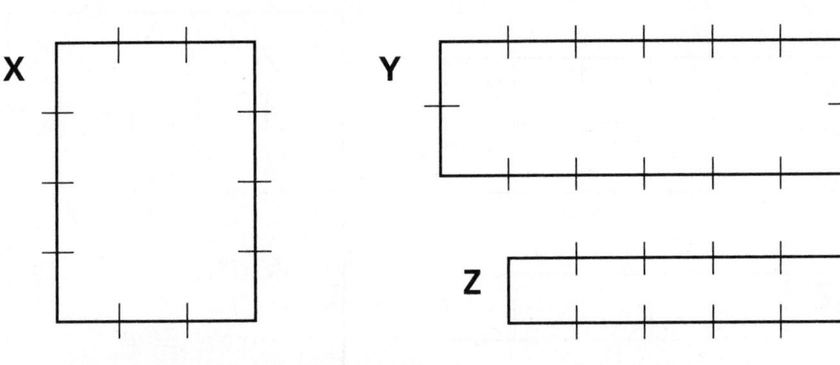

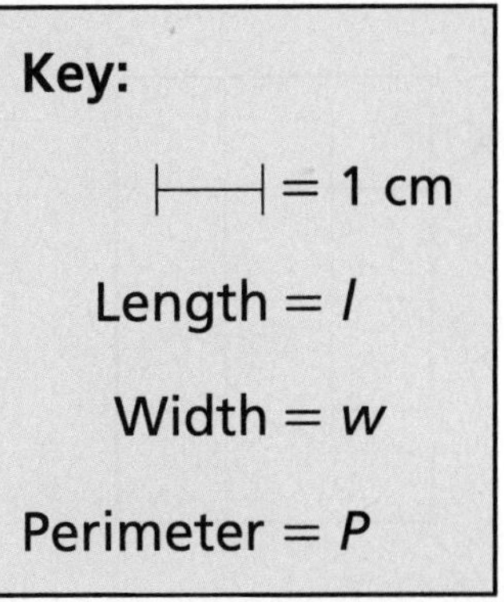

1. The measurement unit for these rectangles is 1 centimeter (1 cm). How can you find the total number of centimeters around the outside of each rectangle?

2. What is the perimeter of rectangle X? of rectangle Y? of rectangle Z?

3. How did you find the perimeter of each rectangle?

4. Look at the key: **length** is the distance across a rectangle and **width** is the distance up-and-down. Perimeter is the total distance around the outside. Use the letters *l*, *w*, and *P* to write a **formula** for the perimeter of rectangles.

VOCABULARY
area
square units

▶ Units of Area

Area is the total number of **square units** that cover a figure. Each square unit inside these rectangles is 1 cm long and 1 cm wide, so it is 1 square centimeter (1 sq cm).

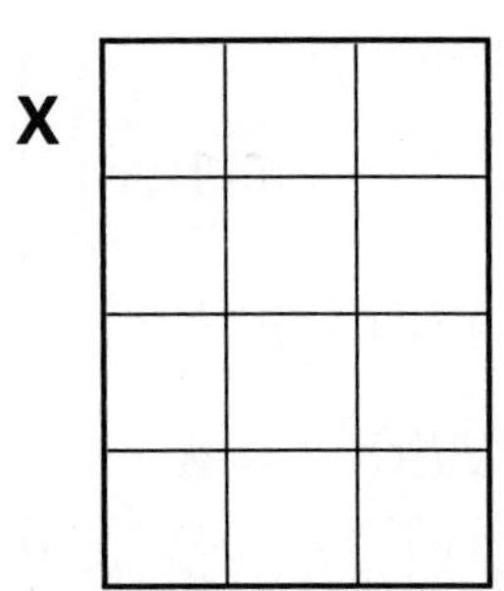

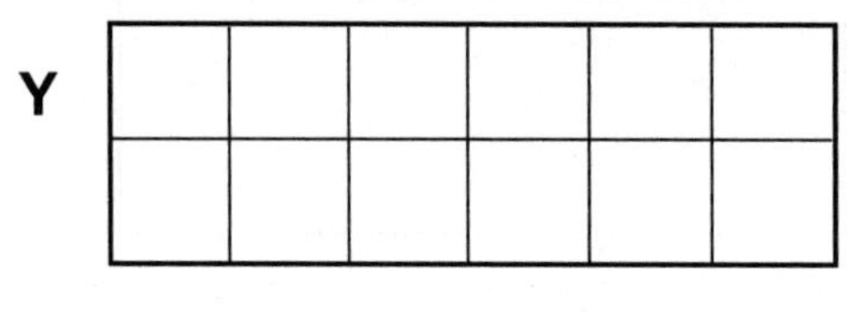

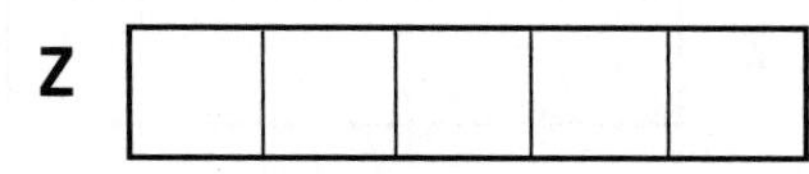

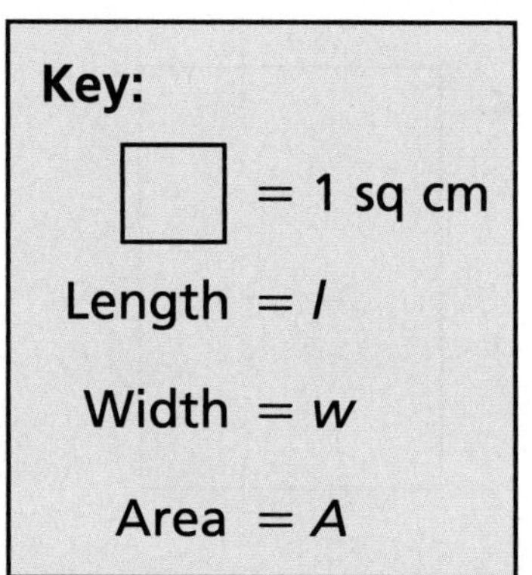

Key:
= 1 sq cm
Length = l
Width = w
Area = A

5. How can you find the total number of square centimeters inside each rectangle?

6. What is the area of rectangle X? of rectangle Y? of rectangle Z?

7. Using l to stand for length, w to stand for width, and A to stand for area, what formula can you write for finding the area of any rectangle?

8. Why does the same formula work for all rectangles?

▶ Review Perimeter and Area

Perimeter and Area

Perimeter and area are measured with different kinds of units: units of distance or length for perimeter and square units for area.

Perimeter is the total distance around the outside of a figure.

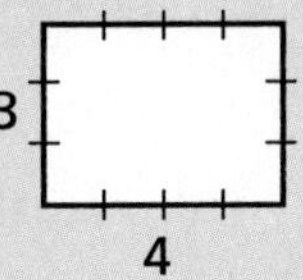

This rectangle has 4 units along its length and 3 units along its width. To find the perimeter, you add the distances of all of the sides:

$$l + w + l + w = P$$

Area is the total number of square units that cover a figure.

For rectangles, area can be seen as an array of squares. This rectangle is an array of 4 squares across (length) and 3 squares down (width). To find its area, you can multiply length times width:

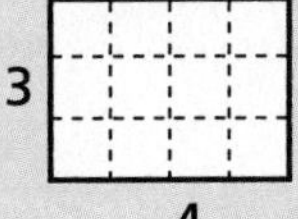

$$l \times w = A$$

▶ Practice with Perimeter and Area

Find the perimeter of rectangle A. Find the area of rectangle B.
1 unit = 1 inch

9. 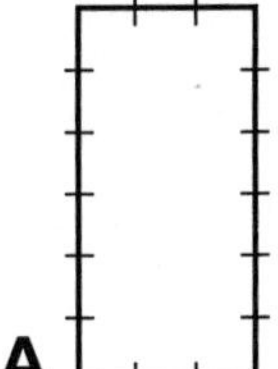

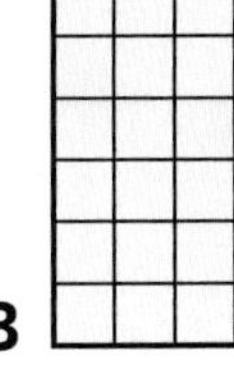

10. 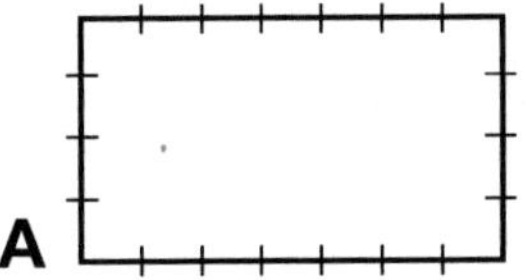

B

Show your work on your paper or in your journal.

▶ Calculate Perimeter and Area

Find the perimeter and area of each rectangle.

11. 9 mi, 2 mi

12. 4 m, 4 m

13. 5 ft, 3 ft

14. 8 cm, 4 cm

Solve.

15. The area of the rectangle is 60 square meters. One side of the rectangle has a length of 10 meters. What is the unknown side length?

10 m

16. The perimeter of the rectangle is 32 inches. One side of the rectangle has a length of 11 inches. What is the unknown side length?

11 in.

17. The perimeter of a rectangle is 56 mm. Three of the sides have a length of 15 mm, 13 mm, and 15 mm. What is the unknown side length?

18. The area of a rectangle is 81 sq inches. One side of the rectangle has a length of 9 in. What is the unknown side length?

5-7 Class Activity

Show your work on your paper or in your journal.

► Solve Real World Measurement Problems

1. Ethan has 6 meters of string. He cuts the string into 3 pieces of equal length. How many centimeters long is each piece of string?

2. Jamal put a cantaloupe with a mass of 450 grams in a bag. He adds another cantaloupe that had a mass of 485 grams. How many grams of cantaloupe are in the bag?

3. Tanisha has 35 milliliters of apple juice. She divides the juice evenly into 5 different glasses. How much juice does she pour into each glass?

4. Katherine had a box of rocks that weighed 4 pounds. She put the rocks into 2 bags, each with the same weight. What is the weight of each bag, in ounces?

5. Grace buys a 2 foot piece of yellow ribbon and an 8 foot piece of pink ribbon at the craft store. How many times as long is the pink ribbon as the yellow ribbon? How many inches of ribbon does Grace have in all?

6. Adriana has one gallon of juice. She pours the juice into containers that each holds one pint of juice. She gives two pints to her friends. How many pints of juice are left?

7. Jamison ran 1,780 feet yesterday. Today he ran 2,165 feet. How many yards did Jamison run the past two days?

▶ Solve Real World Measurement Problems (continued)

8. A supermarket sells 89 gallons of milk in January and 82 gallons of milk in February. Altogether, how many gallons of milk did the supermarket sell in January and February?

Show your work on your paper or in your journal.

9. Felix has a dog that weighs 38 pounds. Felix's dog weighs twice as much as Marcell's dog does. In ounces, how much more does Felix's dog weigh than Marcell's dog?

10. Grant ran 500 meters around a track. Harry ran 724 meters around the same track. How many more meters did Harry run than Grant?

▶ Perimeter and Area Word Problems

11. The area of the rectangular sandbox is 32 square feet. The short side of the sandbox measures 4 feet. How long is the long side of the sandbox?

12. One wall in Dennis's square bedroom is 13 feet long. What is the perimeter of Dennis's bedroom?

13. A square playground has an area of 900 square feet. What is the length of each side of the playground?

14. A rectangular rug has a perimeter of 20 feet. The length of the rug is 6 feet. What is the width of the rug in inches?

▶ Math and Gardens

Gardens come in all shapes and sizes and can include flowers, vegetables, and many other plants. A Dutch garden is a type of garden that is often a rectangle made up of smaller rectangles and squares, called flowerbeds. A hedge or wall is often placed around the perimeter of the garden. Dutch gardens are known for having very colorful, tightly packed flowers. The Sunken Garden is a famous Dutch garden at Kensington Palace in London, England.

Jared looked up some information about taking care of a garden. He discovered that you need to water and fertilize a garden regularly to help the plants grow. The information he found is shown in the table at the right.

Gardening Information	
Fertilizer	4 ounces per 100 sq feet
Water	30 gallons every 3 days

Use the diagram below to answer the questions. It shows a planned flowerbed for a Dutch garden. The perimeter of the flowerbed is 200 feet.

?

10 ft

1. What is the length of the unknown side?

2. What is the area of the garden?

Show your work on your paper or in your journal.

3. How many ounces of fertilizer should be used on the flowerbed?

4. How many cups of water are used every 9 days?

▶ Rectangles in Gardens

Padma wants to create a rectangular shaped garden in her backyard. She wants to have a total of three flowerbeds, two of which will be the same size. She drew a diagram of how she wants the garden to look. Use the diagram to answer the questions below.

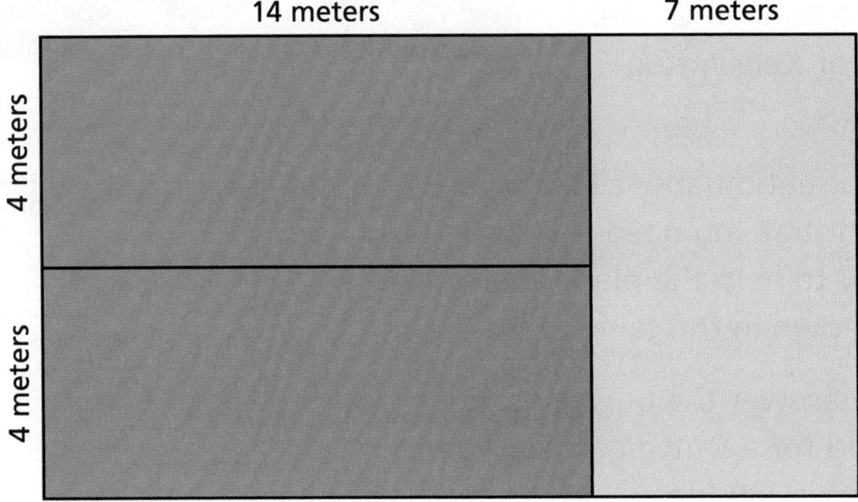

5. What is the perimeter of the entire blue section of flowerbeds?

6. Which section of the garden has a greater perimeter, the green section or the entire blue section? How much greater?

7. What is the area of the whole garden?

8. Padma decides to plant tulips in one of the blue flowerbeds and roses in the green flowerbed. Compare the area of the tulip flowerbed and the rose flowerbed using >, <, or =.

Use the Activity Workbook Unit Test on pages 55–56.

▶ Vocabulary

VOCABULARY
area
metric system
perimeter
pound

Choose the best term from the box.

1. The ________ is the measurement of the distance around the outside of a figure. (Lesson 5-6)
2. One ________ is equal to 16 ounces. (Lesson 5-5)
3. The ________ is the total number of square units that cover a figure. (Lesson 5-6)

▶ Concepts and Skills

4. Explain how to find how many cups are in 8 quarts. (Lesson 5-5)

5. Explain why the formula for the perimeter of a rectangle and the formula for the perimeter of a square are different. (Lesson 5-6)

Convert. (Lessons 5-1, 5-2, 5-3, 5-4, 5-5)

6. 40 m = ________ cm
7. 65 L = ________ cL
8. 3 kg = ________ g
9. 6 yd = ________ ft
10. 3 lb = ________ oz
11. 9 gal = ________ pt
12. 7 hours = ________ min
13. 8 years = ________ months
14. 21 min = ________ sec

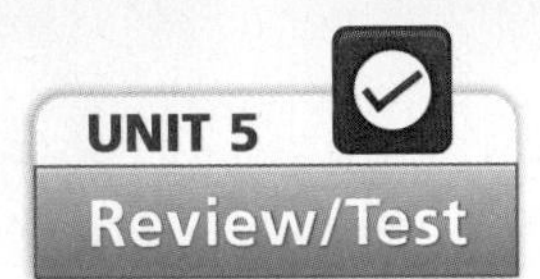

Find the area and perimeter of each rectangle. (Lessons 5-6)

15. 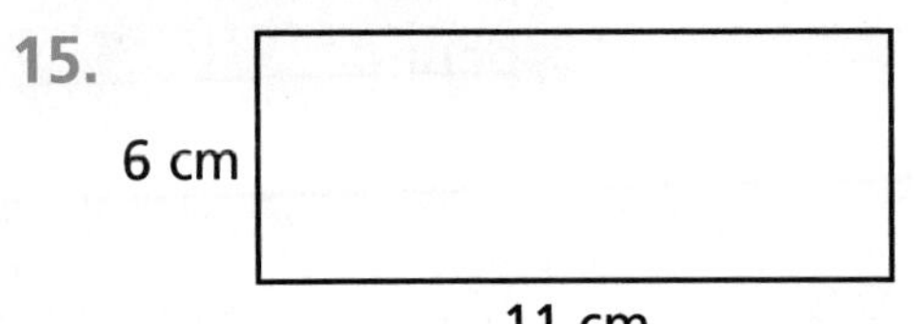

$P = $ ■

$A = $ ■

16. 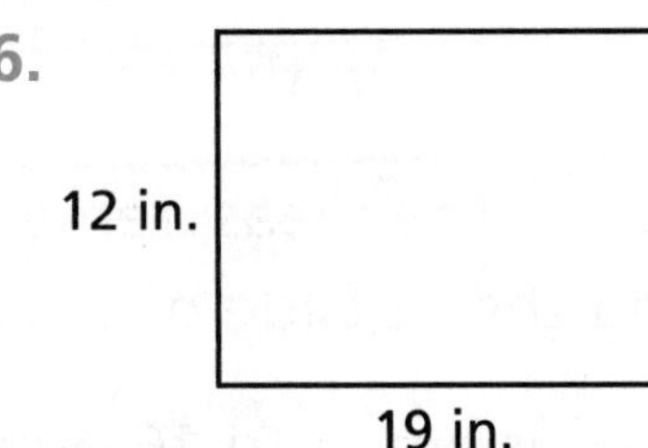

$P = $ ■

$A = $ ■

▶ Problem Solving

Solve.

17. A movie starts at 12:45 P.M. and is exactly 1 hour and 35 minutes long. What time does the movie end? (Lessons 5-3, 5-7)

18. A rectangular kitchen has an area of 126 square feet. The length is 14 feet. What is the width? (Lesson 5-6, 5-7)

19. Angie buys 6 feet of red ribbon and 8 feet of blue ribbon for a project. How many inches of ribbon did Angie buy in all? (Lessons 5-2, 5-7)

20. **Extended Response** Jack buys some rocks. Each rock has a mass of 4 kilograms. He buys 19 rocks. How many grams of rock did Jack buy? Explain how you solve this problem. (Lessons 5-2, 5-7)

Dear Family,

Your child has experience with fractions through measurements and in previous grades. Unit 6 of *Math Expressions* builds on this experience. The main goals of this unit are to:

- understand the meaning of fractions.
- compare unit fractions.
- add and subtract fractions and mixed numbers with like denominators.
- multiply a fraction by a whole number.

Your child will use fraction bars and fraction strips to gain a visual and conceptual understanding of fractions as parts of a whole. Later, your child will use these models to add and subtract fractions and to convert between improper fractions and mixed numbers.

Examples of Fraction Bar Modeling:

Share with your family the Family Letter on Activity Workbook page 57.

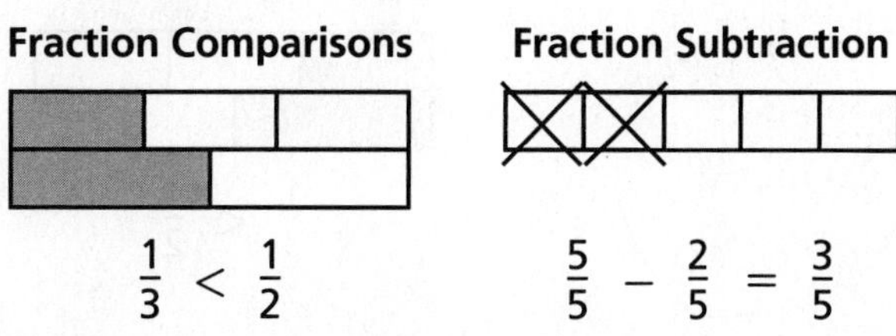

Fraction Comparisons
Fraction Subtraction
$\frac{1}{3} < \frac{1}{2}$
$\frac{5}{5} - \frac{2}{5} = \frac{3}{5}$

In later lessons of this unit, your child will be introduced to the number line model for fractions. Students name fractions corresponding to given lengths on the number line and identify lengths corresponding to given fractions. They also see that there are many equivalent fraction names for any given length.

Your child will apply this knowledge about fractions and fraction operations to solve real world problems.

If you have questions or problems, please contact me.

Sincerely,
Your child's teacher

COMMON CORE — This unit includes the Common Core Standards for Mathematical Content for Numbers and Operations-Fractions, 4.NF.3, 4.NF.3a, 4.NF.3b, 4.NF.3c, 4.NF.3d, 4.NF.4a, 4.NF.4b, 4.NF.4c, and all Mathematical Practices.

Estimada familia:

Su niño ha usado fracciones al hacer mediciones y en los grados previos. La Unidad 6 de *Math Expressions* amplía esta experiencia. Los objetivos principales de la unidad son:

- comprender el significado de las fracciones.
- comparar fracciones unitarias.
- sumar y restar fracciones y números mixtos con denominadores iguales.
- multiplicar una fracción por un número entero.

Su niño usará barras y tiras de fracciones para comprender y visualizar el concepto de las fracciones como partes de un entero. Luego, usará estos modelos para sumar y restar fracciones y para convertir fracciones impropias y números mixtos.

Ejemplos de modelos con barras de fracciones:

Muestra a tu familia la Carta a la familia de la página 58 del Cuaderno de actividades y trabajo.

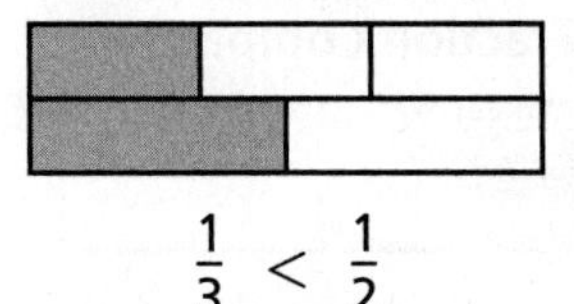

$\frac{1}{3} < \frac{1}{2}$

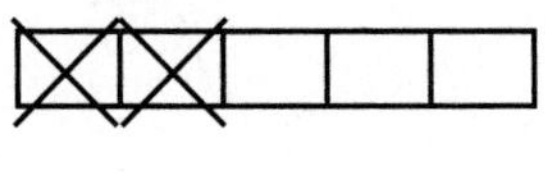

$\frac{5}{5} - \frac{2}{5} = \frac{3}{5}$

Más adelante en esta unidad, su niño verá el modelo de la recta numérica para las fracciones. Los estudiantes nombrarán las fracciones que correspondan a determinadas longitudes en la recta numérica e identificarán longitudes que corresponden a fracciones dadas. También observarán que hay muchos nombres de fracciones equivalentes para una longitud determinada.

Su niño aplicará este conocimiento de las fracciones y operaciones con fracciones para resolver problemas cotidianos.

Si tiene alguna duda o algún comentario, por favor comuníquese conmigo.

Atentamente,
El maestro de su niño

Esta unidad incluye los Common Core Standards for Mathematical Content for Numbers and Operations-Fractions, 4.NF.3, 4.NF.3a, 4.NF.3b, 4.NF.3c, 4.NF.3d, 4.NF.4a, 4.NF.4b, 4.NF.4c, and all Mathematical Practices.

VOCABULARY
unit fraction
fraction
numerator
denominator

► Sums of Fractions

A **unit fraction** represents one equal part of a whole. A unit fraction has a numerator of 1. The unit fraction $\frac{1}{d}$ is one of d equal parts.

The fraction bar below is divided into six equal parts, or sixths. Each part is 1 of 6 equal parts, or $\frac{1}{6}$.

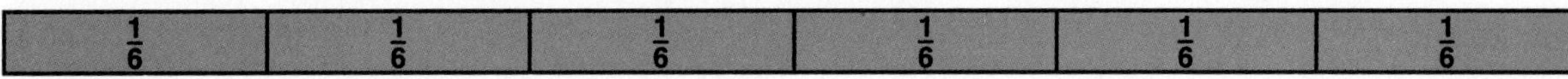

A **fraction** is the sum of unit fractions.
The fraction $\frac{n}{d}$ is the sum of n copies of $\frac{1}{d}$.

numerator → **denominator** →

$$\frac{n}{d} = \frac{\text{number of unit fractions in the fraction}}{\text{number of equal parts in the whole}}$$

The fraction $\frac{5}{6}$ is the sum of five sixths.

$$\frac{5}{6} = \frac{1}{6} + \frac{1}{6} + \frac{1}{6} + \frac{1}{6} + \frac{1}{6} = 5 \times \frac{1}{6}$$

$$\frac{1}{6} + \frac{1}{6} + \frac{1}{6} + \frac{1}{6} + \frac{1}{6} = \frac{5}{6}$$

$\frac{1}{6}$	$\frac{1}{6}$	$\frac{1}{6}$	$\frac{1}{6}$	$\frac{1}{6}$	$\frac{1}{6}$

Fold your fraction strips to show each sum of unit fractions. Write the fraction each sum represents.

1. $\frac{1}{3} + \frac{1}{3} = $ ■

2. $\frac{1}{8} + \frac{1}{8} + \frac{1}{8} + \frac{1}{8} + \frac{1}{8} = $ ■

3. $\frac{1}{4} + \frac{1}{4} = $ ■

4. $\frac{1}{6} + \frac{1}{6} + \frac{1}{6} + \frac{1}{6} = $ ■

5. $\frac{1}{12} + \frac{1}{12} + \frac{1}{12} + \frac{1}{12} + \frac{1}{12} + \frac{1}{12} = $ ■

6. $\frac{1}{12} + \frac{1}{12} + \frac{1}{12} + \frac{1}{12} + \frac{1}{12} + \frac{1}{12} + \frac{1}{12} + \frac{1}{12} = $ ■

7. $\frac{1}{8} + \frac{1}{8} + \frac{1}{8} + \frac{1}{8} + \frac{1}{8} + \frac{1}{8} + \frac{1}{8} = $ ■

► Patterns in Fraction Bars

8. Describe at least three patterns you see in the fraction bars below.

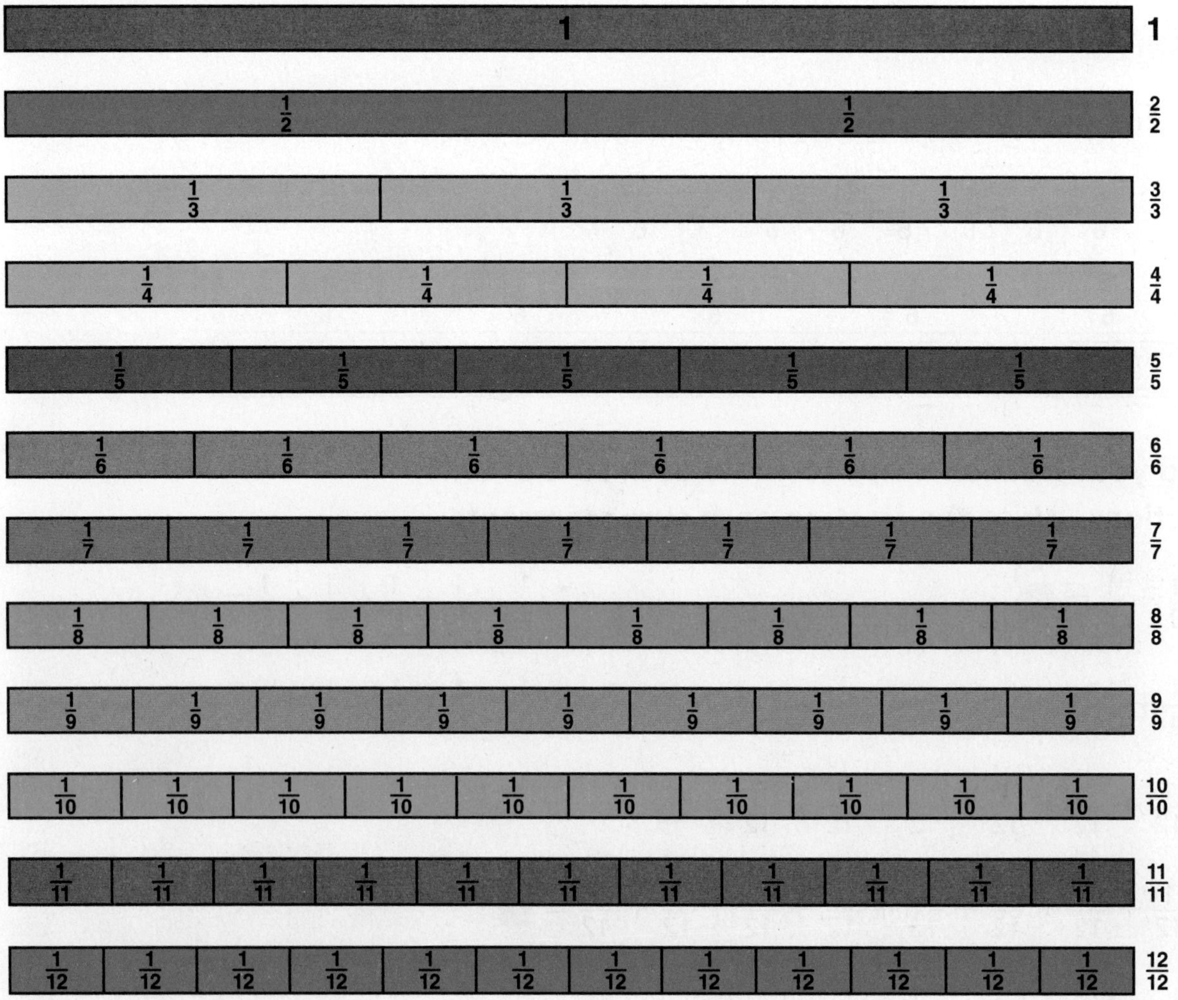

1	1
$\frac{1}{2}$ $\frac{1}{2}$	$\frac{2}{2}$
$\frac{1}{3}$ $\frac{1}{3}$ $\frac{1}{3}$	$\frac{3}{3}$
$\frac{1}{4}$ $\frac{1}{4}$ $\frac{1}{4}$ $\frac{1}{4}$	$\frac{4}{4}$
$\frac{1}{5}$ $\frac{1}{5}$ $\frac{1}{5}$ $\frac{1}{5}$ $\frac{1}{5}$	$\frac{5}{5}$
$\frac{1}{6}$ $\frac{1}{6}$ $\frac{1}{6}$ $\frac{1}{6}$ $\frac{1}{6}$ $\frac{1}{6}$	$\frac{6}{6}$
$\frac{1}{7}$ $\frac{1}{7}$ $\frac{1}{7}$ $\frac{1}{7}$ $\frac{1}{7}$ $\frac{1}{7}$ $\frac{1}{7}$	$\frac{7}{7}$
$\frac{1}{8}$ $\frac{1}{8}$ $\frac{1}{8}$ $\frac{1}{8}$ $\frac{1}{8}$ $\frac{1}{8}$ $\frac{1}{8}$ $\frac{1}{8}$	$\frac{8}{8}$
$\frac{1}{9}$ $\frac{1}{9}$ $\frac{1}{9}$ $\frac{1}{9}$ $\frac{1}{9}$ $\frac{1}{9}$ $\frac{1}{9}$ $\frac{1}{9}$ $\frac{1}{9}$	$\frac{9}{9}$
$\frac{1}{10}$ $\frac{1}{10}$ $\frac{1}{10}$ $\frac{1}{10}$ $\frac{1}{10}$ $\frac{1}{10}$ $\frac{1}{10}$ $\frac{1}{10}$ $\frac{1}{10}$ $\frac{1}{10}$	$\frac{10}{10}$
$\frac{1}{11}$ $\frac{1}{11}$ $\frac{1}{11}$ $\frac{1}{11}$ $\frac{1}{11}$ $\frac{1}{11}$ $\frac{1}{11}$ $\frac{1}{11}$ $\frac{1}{11}$ $\frac{1}{11}$ $\frac{1}{11}$	$\frac{11}{11}$
$\frac{1}{12}$ $\frac{1}{12}$ $\frac{1}{12}$ $\frac{1}{12}$ $\frac{1}{12}$ $\frac{1}{12}$ $\frac{1}{12}$ $\frac{1}{12}$ $\frac{1}{12}$ $\frac{1}{12}$ $\frac{1}{12}$ $\frac{1}{12}$	$\frac{12}{12}$

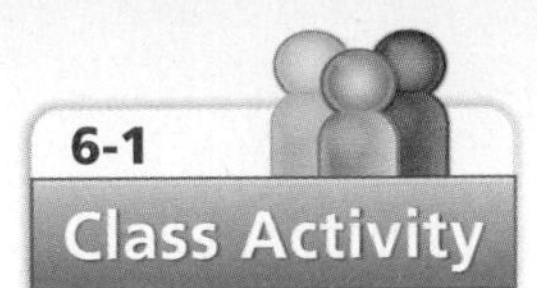

Use Activity Workbook page 59.

▶ Sums of Unit Fractions

Shade the fraction bar to show each fraction. Then write the fraction as a sum of unit fractions and as a product of a whole number and a unit fraction. The first one is done for you.

9. $\frac{3}{4} = \underline{\frac{1}{4} + \frac{1}{4} + \frac{1}{4}} = \underline{3 \times \frac{1}{4}}$

$\frac{1}{4}$ (shaded)	$\frac{1}{4}$ (shaded)	$\frac{1}{4}$ (shaded)	$\frac{1}{4}$

10. $\frac{3}{8}$ = ______ = ______

$\frac{1}{8}$	$\frac{1}{8}$	$\frac{1}{8}$	$\frac{1}{8}$	$\frac{1}{8}$	$\frac{1}{8}$	$\frac{1}{8}$	$\frac{1}{8}$

11. $\frac{5}{5}$ = ______ = ______

$\frac{1}{5}$	$\frac{1}{5}$	$\frac{1}{5}$	$\frac{1}{5}$	$\frac{1}{5}$

12. $\frac{2}{12}$ = ______ = ______

$\frac{1}{12}$	$\frac{1}{12}$	$\frac{1}{12}$	$\frac{1}{12}$	$\frac{1}{12}$	$\frac{1}{12}$	$\frac{1}{12}$	$\frac{1}{12}$	$\frac{1}{12}$	$\frac{1}{12}$	$\frac{1}{12}$	$\frac{1}{12}$

13. $\frac{4}{7}$ = ______ = ______

$\frac{1}{7}$	$\frac{1}{7}$	$\frac{1}{7}$	$\frac{1}{7}$	$\frac{1}{7}$	$\frac{1}{7}$	$\frac{1}{7}$

14. $\frac{7}{9}$ = ______ = ______

$\frac{1}{9}$	$\frac{1}{9}$	$\frac{1}{9}$	$\frac{1}{9}$	$\frac{1}{9}$	$\frac{1}{9}$	$\frac{1}{9}$	$\frac{1}{9}$	$\frac{1}{9}$

► Fractions as Parts of a Whole

Jon made a large sandwich for the 6 people in his family. He asked his father to help him cut it into 6 equal pieces. To do this, they made a paper cutting guide that is as long as the sandwich. Jon folded the paper into 6 equal parts, and his father used it to cut the sandwich into equal pieces.

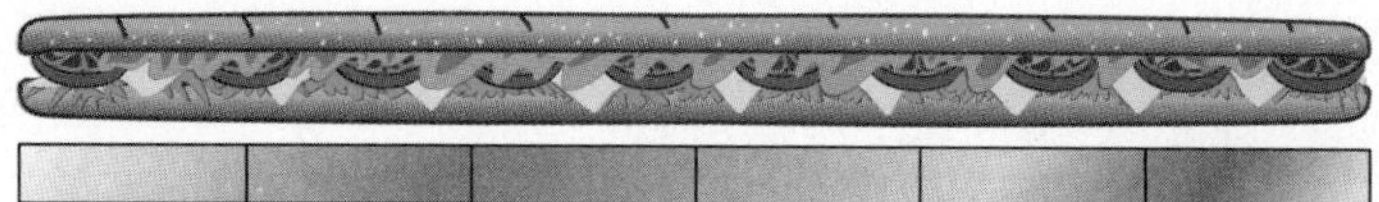

Show your work on your paper or in your journal.

Solve.

15. If each person ate 1 piece of the sandwich, what fraction of the sandwich did each person eat? Fold your 6-part fraction strip to show the fraction of the whole sandwich that each person ate.

16. How many pieces of the whole sandwich did Jon's mother and father eat altogether? Fold your fraction strip to show the fraction of the whole sandwich Jon's mother and father ate in all.

17. After Jon's mother and father got their pieces, what fraction of the sandwich was left?

18. Jon and each of his sisters were each able to have one piece of the remaining sandwiches. How many sisters does Jon have?

19. What ideas about fractions did we use to answer the questions about Jon's sandwich?

6-2
Class Activity

Use Activity
Workbook page 60.

► Fifths that Add to One

Every afternoon, student volunteers help the school librarian put returned books back on the shelves. The librarian puts the books in equal piles on a cart.

One day, Jean and Maria found 5 equal piles on the return cart. They knew there were different ways they could share the job of reshelving the books. They drew fraction bars to help them find all the possibilities.

1. On each fifths bar, circle two groups of fifths to show one way Jean and Maria could share the work. (Each bar should show a different possibility.) Then complete the equation next to each bar to show their shares.

1 whole = all of the books				
$\frac{1}{5}$	$\frac{1}{5}$	$\frac{1}{5}$	$\frac{1}{5}$	$\frac{1}{5}$

1 whole = Jean's share + Maria's share

$\frac{5}{5} = \frac{\square}{5} + \frac{\square}{5}$

$\frac{1}{5}$	$\frac{1}{5}$	$\frac{1}{5}$	$\frac{1}{5}$	$\frac{1}{5}$

$\frac{5}{5} = \frac{\square}{5} + \frac{\square}{5}$

$\frac{1}{5}$	$\frac{1}{5}$	$\frac{1}{5}$	$\frac{1}{5}$	$\frac{1}{5}$

$\frac{5}{5} = \frac{\square}{5} + \frac{\square}{5}$

$\frac{1}{5}$	$\frac{1}{5}$	$\frac{1}{5}$	$\frac{1}{5}$	$\frac{1}{5}$

$\frac{5}{5} = \frac{\square}{5} + \frac{\square}{5}$

6-2

Class Activity

Use Activity Workbook page 61.

► Sixths that Add to One

The librarian put 6 equal piles of returned books on the cart for Liu and Henry to reshelve. They also drew fraction bars.

2. On each sixths bar, circle two groups of sixths to show one way that Liu and Henry could share the work. (Each bar should show a different possibility.) Then complete the equation next to each bar to show their shares.

1 whole = all of the books						1 whole		Liu's share		Henry's share
$\frac{1}{6}$	$\frac{1}{6}$	$\frac{1}{6}$	$\frac{1}{6}$	$\frac{1}{6}$	$\frac{1}{6}$	$\frac{6}{6}$	=	$\frac{■}{6}$	+	$\frac{■}{6}$
$\frac{1}{6}$	$\frac{1}{6}$	$\frac{1}{6}$	$\frac{1}{6}$	$\frac{1}{6}$	$\frac{1}{6}$	$\frac{6}{6}$	=	$\frac{■}{6}$	+	$\frac{■}{6}$
$\frac{1}{6}$	$\frac{1}{6}$	$\frac{1}{6}$	$\frac{1}{6}$	$\frac{1}{6}$	$\frac{1}{6}$	$\frac{6}{6}$	=	$\frac{■}{6}$	+	$\frac{■}{6}$
$\frac{1}{6}$	$\frac{1}{6}$	$\frac{1}{6}$	$\frac{1}{6}$	$\frac{1}{6}$	$\frac{1}{6}$	$\frac{6}{6}$	=	$\frac{■}{6}$	+	$\frac{■}{6}$
$\frac{1}{6}$	$\frac{1}{6}$	$\frac{1}{6}$	$\frac{1}{6}$	$\frac{1}{6}$	$\frac{1}{6}$	$\frac{6}{6}$	=	$\frac{■}{6}$	+	$\frac{■}{6}$

► Find the Unknown Addend

Write the fraction that will complete each equation.

3. $1 = \frac{7}{7} = \frac{1}{7} + ■$

4. $1 = \frac{4}{4} = \frac{3}{4} + ■$

5. $1 = \frac{8}{8} = \frac{3}{8} + ■$

6. $1 = \frac{5}{5} = \frac{2}{5} + ■$

7. $1 = \frac{3}{3} = \frac{2}{3} + ■$

8. $1 = \frac{10}{10} = \frac{6}{10} + ■$

9. $1 = \frac{6}{6} = \frac{2}{6} + ■$

10. $1 = \frac{8}{8} = \frac{5}{8} + ■$

► Discuss and Compare Unit Fractions

Use these fraction bars to help you compare the unit fractions. Write > or <.

1												$\frac{1}{1}$
$\frac{1}{2}$	$\frac{1}{2}$											$\frac{2}{2}$
$\frac{1}{3}$	$\frac{1}{3}$	$\frac{1}{3}$										$\frac{3}{3}$
$\frac{1}{4}$	$\frac{1}{4}$	$\frac{1}{4}$	$\frac{1}{4}$									$\frac{4}{4}$
$\frac{1}{5}$	$\frac{1}{5}$	$\frac{1}{5}$	$\frac{1}{5}$	$\frac{1}{5}$								$\frac{5}{5}$
$\frac{1}{6}$	$\frac{1}{6}$	$\frac{1}{6}$	$\frac{1}{6}$	$\frac{1}{6}$	$\frac{1}{6}$							$\frac{6}{6}$
$\frac{1}{7}$	$\frac{1}{7}$	$\frac{1}{7}$	$\frac{1}{7}$	$\frac{1}{7}$	$\frac{1}{7}$	$\frac{1}{7}$						$\frac{7}{7}$
$\frac{1}{8}$	$\frac{1}{8}$	$\frac{1}{8}$	$\frac{1}{8}$	$\frac{1}{8}$	$\frac{1}{8}$	$\frac{1}{8}$	$\frac{1}{8}$					$\frac{8}{8}$
$\frac{1}{9}$	$\frac{1}{9}$	$\frac{1}{9}$	$\frac{1}{9}$	$\frac{1}{9}$	$\frac{1}{9}$	$\frac{1}{9}$	$\frac{1}{9}$	$\frac{1}{9}$				$\frac{9}{9}$
$\frac{1}{10}$	$\frac{1}{10}$	$\frac{1}{10}$	$\frac{1}{10}$	$\frac{1}{10}$	$\frac{1}{10}$	$\frac{1}{10}$	$\frac{1}{10}$	$\frac{1}{10}$	$\frac{1}{10}$			$\frac{10}{10}$
$\frac{1}{11}$	$\frac{1}{11}$	$\frac{1}{11}$	$\frac{1}{11}$	$\frac{1}{11}$	$\frac{1}{11}$	$\frac{1}{11}$	$\frac{1}{11}$	$\frac{1}{11}$	$\frac{1}{11}$	$\frac{1}{11}$		$\frac{11}{11}$
$\frac{1}{12}$	$\frac{1}{12}$	$\frac{1}{12}$	$\frac{1}{12}$	$\frac{1}{12}$	$\frac{1}{12}$	$\frac{1}{12}$	$\frac{1}{12}$	$\frac{1}{12}$	$\frac{1}{12}$	$\frac{1}{12}$	$\frac{1}{12}$	$\frac{12}{12}$

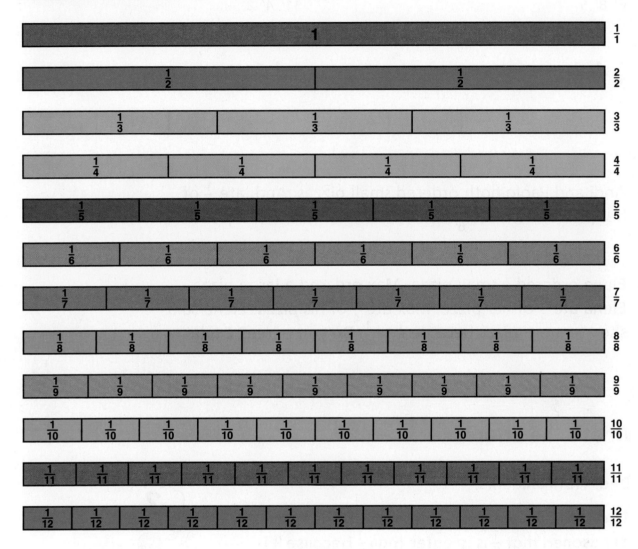

11. $\frac{1}{6} \bigcirc \frac{1}{8}$

12. $\frac{1}{5} \bigcirc \frac{1}{3}$

13. $\frac{1}{10} \bigcirc \frac{1}{12}$

14. $\frac{1}{7} \bigcirc \frac{1}{4}$

15. $\frac{1}{9} \bigcirc \frac{1}{12}$

16. $\frac{1}{9} \bigcirc \frac{1}{11}$

17. Complete this statement:

    When comparing two unit fractions, the fraction with the smaller denominator is ______.

Show your work on your paper or in your journal.

► Compare and Order Unit Fractions

Write the unit fractions in order from least to greatest.

18. $\frac{1}{6}, \frac{1}{8}, \frac{1}{5}$

19. $\frac{1}{11}, \frac{1}{4}, \frac{1}{8}$

20. $\frac{1}{3}, \frac{1}{10}, \frac{1}{7}$

21. $\frac{1}{4}, \frac{1}{7}, \frac{1}{9}$

Solve

22. Andi and Paolo both ordered small pizzas. Andi ate $\frac{1}{4}$ of her pizza. Paolo ate $\frac{1}{6}$ of his pizza. Who ate more pizza?

23. Elena ordered a small pizza. Max ordered a large pizza. Elena ate $\frac{1}{3}$ of her pizza. Max ate $\frac{1}{4}$ of his pizza. Elena said she ate more pizza because $\frac{1}{3} > \frac{1}{4}$. Do you agree? Explain.

► What's the Error?

Dear Math Students,

I had to compare $\frac{1}{4}$ and $\frac{1}{2}$ on my math homework. I reasoned that $\frac{1}{4}$ is greater than $\frac{1}{2}$ because 4 is greater than 2. My friend told me this isn't right. Can you help me understand why my reasoning is wrong?

Your friend,
Puzzled Penguin

24. Write a response to Puzzled Penguin.

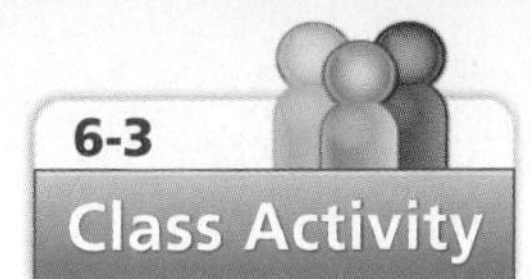

Use Activity Workbook page 62.

▶ Add Fractions

The circled parts of this fraction bar show an addition problem.

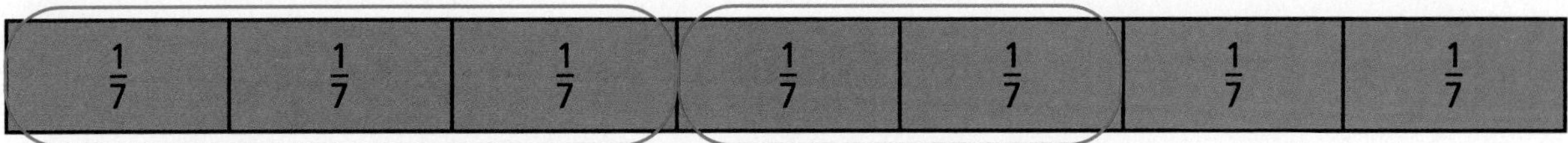

1. Write the numerators that will complete the addition equation.

$\frac{\square}{7} + \frac{\square}{7} = \frac{\square + \square}{7} = \frac{\square}{7}$

Solve each problem. Write the correct numerator to complete each equation.

2. $\frac{3}{9} + \frac{4}{9} = \frac{\square + \square}{9} = \frac{\square}{9}$

3. $\frac{1}{5} + \frac{3}{5} = \frac{\square + \square}{5} = \frac{\square}{5}$

4. $\frac{2}{8} + \frac{5}{8} = \frac{\square + \square}{8} = \frac{\square}{8}$

5. What happens to the numerators in each problem?

6. What happens to the denominators in each problem?

▶ Subtract Fractions

The circled and crossed-out parts of this fraction bar show a subtraction problem.

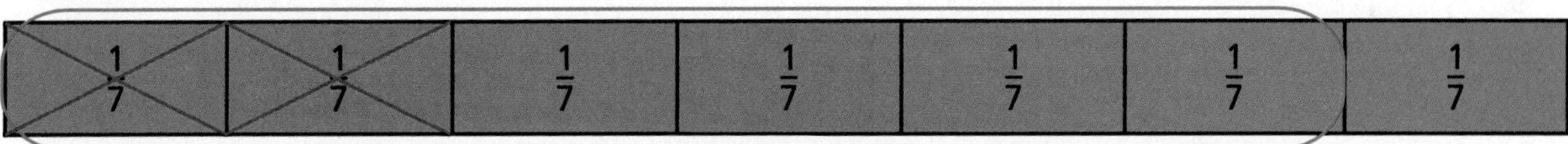

7. Write the numerators that will complete the subtraction equation.

$\frac{\square}{7} - \frac{\square}{7} = \frac{\square - \square}{7} = \frac{\square}{7}$

► Subtract Fractions (continued)

Solve each problem. Write the correct numerators to complete each sentence.

8. $\frac{5}{6} - \frac{4}{6} = \frac{■ - ■}{6} = \frac{■}{6}$

9. $\frac{9}{10} - \frac{5}{10} = \frac{■ - ■}{10} = \frac{■}{■}$

10. $\frac{14}{16} - \frac{9}{16} = \frac{■ - ■}{16} = \frac{■}{■}$

11. What happens to the numerators in each problem?

12. How is subtracting fractions with like denominators similar to adding fractions with like denominators?

► Mixed Practice with Addition and Subtraction

Solve each problem. Include the "circled" step in Exercises 16–21.

13. $\frac{1}{4} + \frac{2}{4} = \frac{■ + ■}{4} = ■$

14. $\frac{3}{9} + \frac{5}{9} = \frac{■ + ■}{9} = ■$

15. $\frac{6}{6} - \frac{2}{6} = \frac{■ - ■}{6} = ■$

16. $\frac{4}{10} + \frac{5}{10} = ■$

17. $\frac{2}{5} + \frac{4}{5} = ■$

18. $\frac{8}{12} - \frac{3}{12} = ■$

19. $\frac{5}{7} + \frac{2}{7} = ■$

20. $\frac{7}{11} - \frac{4}{11} = ■$

21. $\frac{8}{8} - \frac{5}{8} = ■$

Solve.

22. $\frac{7}{9} - \frac{5}{9}$

23. $\frac{4}{5} - \frac{3}{5}$

24. $\frac{1}{3} + \frac{2}{3}$

25. $\frac{2}{11} + \frac{7}{11}$

26. $\frac{5}{6} - \frac{1}{6}$

27. $\frac{1}{8} + \frac{1}{8}$

▶ What's the Error?

Dear Math Students,

My friend said, "If you catch 3 fish and then 2 more fish, how many fish will you have?" Of course, I know I will have 5 fish! She said, "This is the same problem, but you have fifths instead of fish!"

Can you help me understand what my friend meant and help me find the right answer?

Your friend,
Puzzled Penguin

28. Write a response to Puzzled Penguin.

Dear Math Students,

My friend said my answer for this problem is wrong too.

She said "Think about fish again. If you have 4 fish and then eat 3, how many will you have?" What does she mean? What should the answer be?

Your friend,
Puzzled Penguin

29. Write a response to Puzzled Penguin.

6-3
Class Activity

Show your
work on your paper
or in your journal.

► Real World Problems

Draw a model. Then solve.

30. Wayne had $\frac{7}{8}$ cup of trail mix. He ate $\frac{3}{8}$ cup as he was hiking. How many cups does he have now?

31. Reese had $\frac{2}{4}$ cup of orange juice. She added pineapple juice to make a total of $\frac{3}{4}$ cup of juice. How much pineapple juice did she add?

Write an equation. Then solve.

32. Nasira walks $\frac{4}{5}$ mile to school each day. This is $\frac{2}{5}$ mile farther than Kat walks. How far does Kat walk to school?

33. A puppy is now 5 weeks old. It has gained $\frac{8}{16}$ pound since it was born. The puppy weighs $\frac{11}{16}$ pound now. How much did the puppy weigh when it was born?

34. The water in a tub was $\frac{7}{12}$ foot deep. Then Dom added water until it was $\frac{4}{12}$ foot deeper. How deep is the water now?

35. Jesse had some flour. She used $\frac{3}{4}$ cup in a recipe and had $\frac{1}{4}$ cup of flour left. How much flour did she have to start with?

VOCABULARY
mixed number

▶ Mixed Numbers in the Real World

A **mixed number** is a number that consists of a whole number and a fraction.	$1\frac{4}{6}$ $3\frac{4}{5}$
A fraction greater than 1 has a numerator greater than its denominator.	$\frac{10}{6}$ $\frac{19}{5}$

Mellie's Deli makes sandwiches. This is the price list.

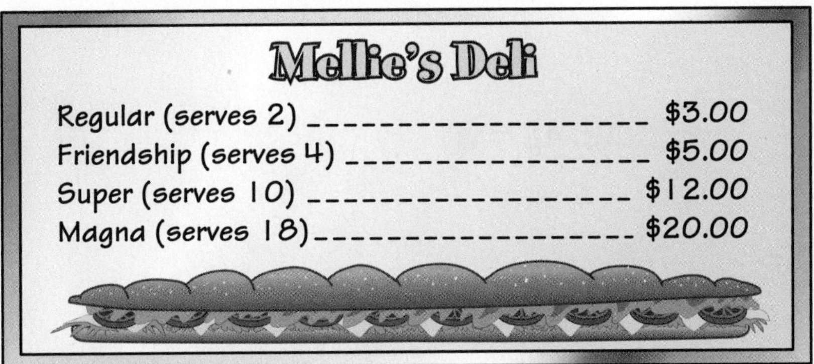

Mellie's Deli
Regular (serves 2) ____________________ $3.00
Friendship (serves 4) _________________ $5.00
Super (serves 10) ____________________ $12.00
Magna (serves 18) ____________________ $20.00

Nineteen friends decide to camp in the park. They order two Super sandwiches. Each camper eats 1 serving.

Solve.

1. How many campers does one Super sandwich serve?

2. What fraction of the second sandwich is needed to serve the rest of the campers?

3. What fraction of the second sandwich is left over?

4. What number tells how many Super sandwiches the campers ate in all?

▶ Convert Between Mixed Numbers and Fractions Greater Than 1

Change each mixed number to a fraction and each fraction to a mixed number.

5. $5\frac{2}{3} = \blacksquare$

6. $3\frac{3}{7} = \blacksquare$

7. $6\frac{6}{10} = \blacksquare$

8. $9\frac{1}{4} = \blacksquare$

9. $2\frac{7}{8} = \blacksquare$

10. $4\frac{5}{9} = \blacksquare$

11. $8\frac{3}{5} = \blacksquare$

12. $7\frac{4}{6} = \blacksquare$

13. $\frac{40}{6} = \blacksquare$

14. $\frac{11}{2} = \blacksquare$

15. $\frac{23}{7} = \blacksquare$

16. $\frac{28}{3} = \blacksquare$

17. $\frac{22}{4} = \blacksquare$

18. $\frac{25}{8} = \blacksquare$

19. $\frac{29}{7} = \blacksquare$

20. $6\frac{4}{8} = \blacksquare$

21. $4\frac{6}{9} = \blacksquare$

22. $\frac{16}{3} = \blacksquare$

Use Activity Workbook page 67.

► Practice Addition and Subtraction with Fractions Greater Than 1

Add or subtract.

1. $\frac{8}{5} + \frac{3}{5} = ■$

2. $\frac{6}{9} + \frac{12}{9} = ■$

3. $\frac{10}{7} - \frac{3}{7} = ■$

4. $\frac{10}{8} + \frac{7}{8} = ■$

5. $\frac{9}{6} - \frac{4}{6} = ■$

6. $\frac{19}{10} - \frac{7}{10} = ■$

► Add Mixed Numbers with Like Denominators

Add.

7. $2\frac{3}{5} + 1\frac{1}{5}$

8. $1\frac{2}{5} + 3\frac{4}{5}$

9. $3\frac{5}{8} + 1\frac{3}{8}$

10. $5\frac{2}{3} + 2\frac{2}{3}$

► Subtract Mixed Numbers with Like Denominators

Subtract.

11. $5\frac{6}{8} - 3\frac{3}{8}$

12. $6\frac{2}{8} - 4\frac{5}{8}$

13. $4\frac{1}{5} - 1\frac{3}{5}$

14. $5\frac{1}{6} - 3\frac{4}{6}$

Explain each solution.

15. $7 + 2 = 9$; $\overset{5}{\cancel{6}}\frac{2}{7} = 5\frac{9}{7}$; $-\ 1\frac{5}{7} = 1\frac{5}{7}$; $4\frac{4}{7}$

16. $6 + 2 = 8$; $\overset{5}{\cancel{6}}\frac{2}{6} = 5\frac{8}{6}$; $-\ 1\frac{5}{6} = 1\frac{5}{6}$; $4\frac{3}{6}$

17. $11 + 2 = 13$; $\overset{5}{\cancel{6}}\frac{2}{11} = 5\frac{13}{11}$; $-\ 1\frac{5}{11} = 1\frac{5}{11}$; $4\frac{8}{11}$

▶ What's the Error?

Dear Math Students,

Here is a subtraction problem that I tried to solve.

Is my answer correct? If not, please help me understand why it is wrong.

Your friend,
Puzzled Penguin

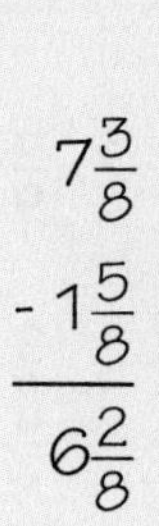

18. Write a response to Puzzled Penguin.

▶ Compare and Subtract

Compare each pair of mixed numbers using $>$ or $<$. Then subtract the lesser mixed number from the greater mixed number.

19. $3\frac{2}{5}$; $1\frac{4}{5}$

20. $\frac{8}{9}$; $2\frac{2}{9}$

21. $\frac{14}{11}$; $1\frac{6}{11}$

22. $4\frac{1}{8}$; $2\frac{7}{8}$

23. $3\frac{2}{6}$; $4\frac{3}{6}$

24. $10\frac{1}{3}$; $7\frac{2}{3}$

▶ Practice with Fractions and Mixed Numbers

Write the fraction that will complete each equation.

1. $1 = \frac{4}{4} = \frac{1}{4} + ■$

2. $1 = \frac{10}{10} = \frac{9}{10} + ■$

3. $1 = \frac{8}{8} = \frac{4}{8} + ■$

Write each fraction as a sum of fractions in two different ways.

4. $\frac{5}{6}$

5. $\frac{8}{10}$

6. $\frac{6}{8}$

7. $\frac{10}{6}$

Write each fraction as a mixed number.

8. $\frac{11}{8} = ■$

9. $\frac{15}{6} = ■$

10. $\frac{32}{5} = ■$

Write each mixed number as a fraction.

11. $3\frac{2}{5} = ■$

12. $1\frac{1}{4} = ■$

13. $2\frac{11}{12} = ■$

Add or subtract.

14. $\frac{2}{5} + \frac{1}{5} = ■$

15. $\frac{3}{9} + \frac{6}{9} = ■$

16. $\frac{4}{6} - \frac{3}{6} = ■$

17. $\frac{5}{7} - \frac{2}{7} = ■$

18. $\frac{7}{12} + \frac{1}{12} = ■$

19. $\frac{10}{10} - \frac{4}{10} = ■$

20. $\frac{9}{4} + \frac{5}{4} = ■$

21. $\frac{23}{8} - \frac{12}{8} = ■$

22. $\frac{5}{2} + \frac{3}{2} = ■$

▶ Practice with Fractions and Mixed Numbers (continued)

Add or subtract.

23. $3\frac{1}{4} + 5\frac{2}{4}$

24. $4\frac{6}{8} - 3\frac{3}{8}$

25. $1\frac{3}{5} + 1\frac{2}{5}$

26. $4\frac{1}{3} - 1\frac{2}{3}$

27. $2\frac{5}{10} + 4\frac{9}{10}$

28. $10\frac{5}{8} - 3\frac{7}{8}$

▶ What's the Error?

Dear Math Students,

This is a problem from my math homework. My friend says my answer is not correct, but I can't figure out what I did wrong. Can you help me find and fix my mistake?

Your friend,
Puzzled Penguin

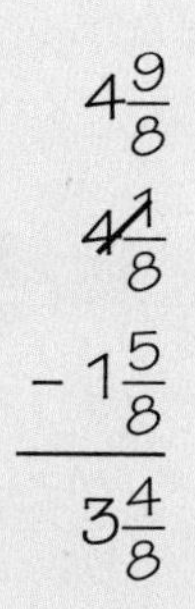

29. Write a response to Puzzled Penguin.

6-6
Class Activity

Show your
work on your paper
or in your journal.

▶ Real World Problems

Write an equation. Then solve.

30. Daniel spent $1\frac{1}{4}$ hours playing soccer on Saturday and $\frac{3}{4}$ hour playing soccer on Sunday. How much time did he spend playing soccer over the weekend?

31. A pitcher contains $4\frac{3}{8}$ cups of juice. Antonio pours $\frac{5}{8}$ cup into a glass. How much juice is left in the pitcher?

32. Shayna walked from school to the library. Then she walked $1\frac{3}{10}$ miles from the library to her apartment. If she walked $2\frac{1}{10}$ miles in all, how far did she walk from school to the library?

33. The vet said Lucy's cat Mittens weighs $7\frac{1}{4}$ pounds. This is $1\frac{2}{4}$ pounds more than Mittens weighed last year. How much did Mittens weigh last year?

34. The width of a rectangle is $3\frac{5}{6}$ inches. The length of the rectangle is $1\frac{4}{6}$ inches longer than the width. What is the length of the rectangle?

35. Choose one of the problems on this page. Draw a model to show that your answer is correct.

Use Activity Workbook page 68.

▶ Make a Line Plot

36. Make a mark anywhere on this line segment.

37. Measure the distance from the left end of the segment to your mark to the nearest quarter inch.

38. Collect measurements from your classmates and record them in the line plot below.

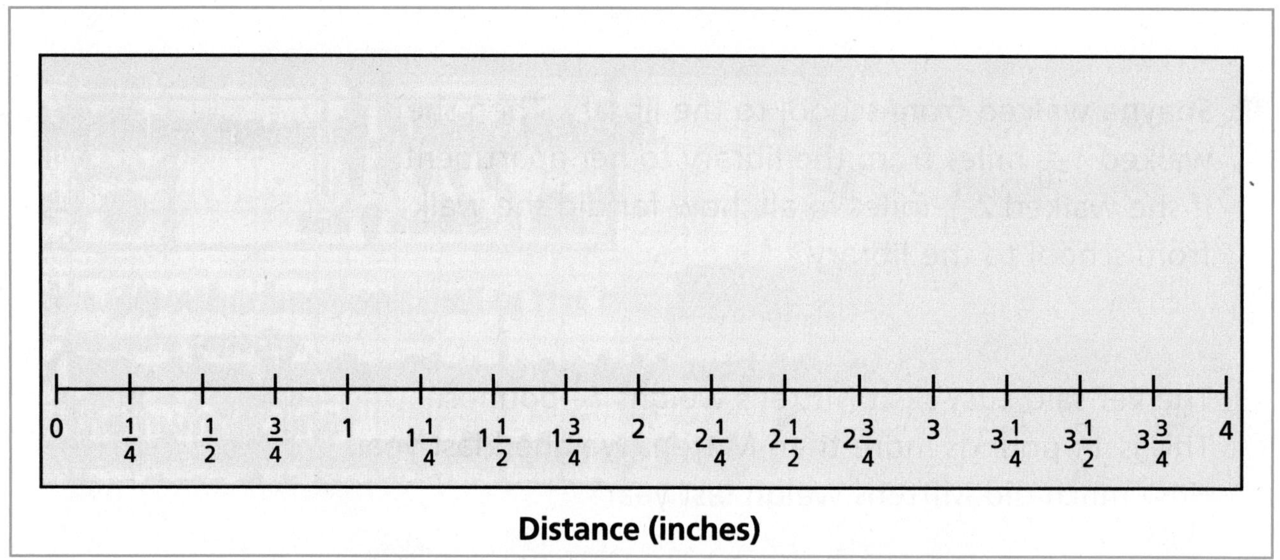

39. The range is the difference between the greatest value and the least value. What is the range of the data?

40. Which distance value was most common?

41. Describe any interesting patterns in the data values. For example, are there any large gaps? Are there clusters of values?

▶ A Whole Number Multiplied by a Unit Fraction

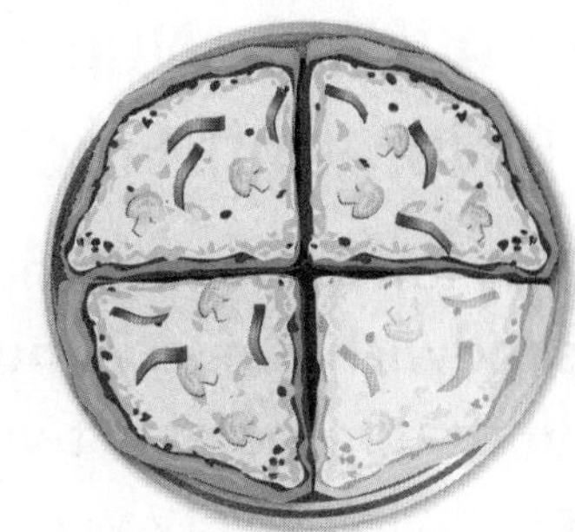

The lunchroom at Mandy's school serves pizza every Friday. Each slice is $\frac{1}{4}$ of a pizza. Mandy eats one slice every week.

To find the fraction of a pizza she eats in three weeks, you can add or multiply.

$$\frac{1}{4} + \frac{1}{4} + \frac{1}{4} = \frac{3}{4} \quad \text{or} \quad 3 \cdot \frac{1}{4} = \frac{3}{4}$$

Solve each problem, first by adding and then by multiplying. Show your work.

1. What fraction of a pizza does she eat in five weeks?

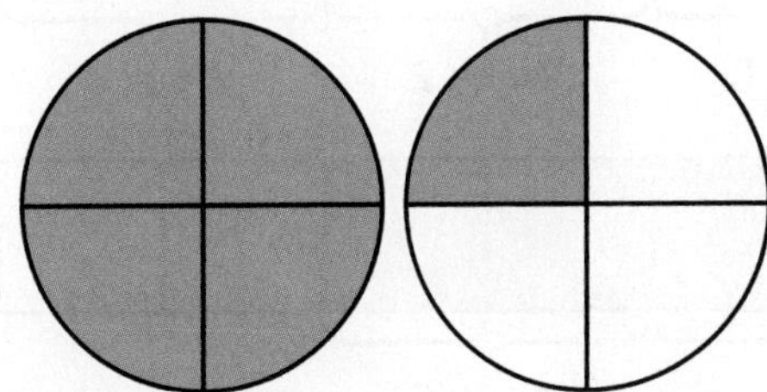

2. What fraction of a pizza does she eat in eleven weeks?

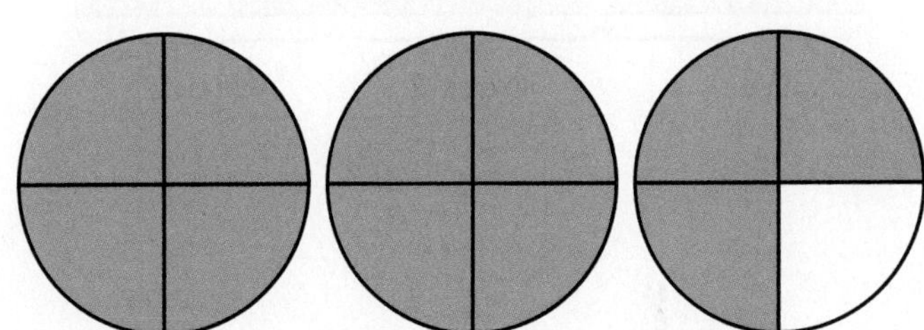

Draw a model for each problem. Then solve.

3. $2 \cdot \frac{1}{3} = ■$

4. $6 \cdot \frac{1}{5} = ■$

5. $10 \cdot \frac{1}{8} = ■$

Draw a model for each fraction. Then write each fraction as the product of a whole number and a unit fraction.

6. $\frac{3}{5} =$ ___■___ $\cdot \frac{1}{5}$

7. $\frac{8}{3} =$ ___■___ $\cdot$ ___■___

8. $\frac{12}{7} =$ ___■___ $\cdot$ ___■___

► A Whole Number Multiplied by a Non-Unit Fraction

The lunchroom at Joe's school serves sub sandwiches every Thursday. Each slice is $\frac{1}{6}$ of a sub. Joe eats two pieces, or $\frac{2}{6}$ of a sandwich, every week.

$\frac{1}{6} + \frac{1}{6} = \frac{2}{6}$ or $2 \cdot \frac{1}{6} = \frac{2}{6}$

Solve each problem, first by adding and then by multiplying. Write your answer as a fraction. Show your work.

9. What fraction of a sandwich does Joe eat in three weeks?

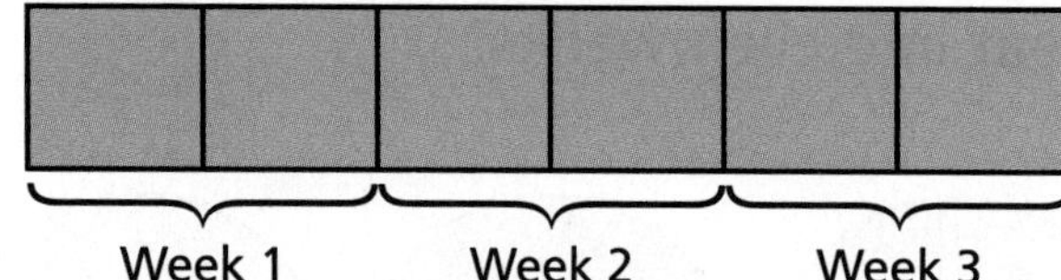

10. What fraction of a sandwich does Joe eat in five weeks?

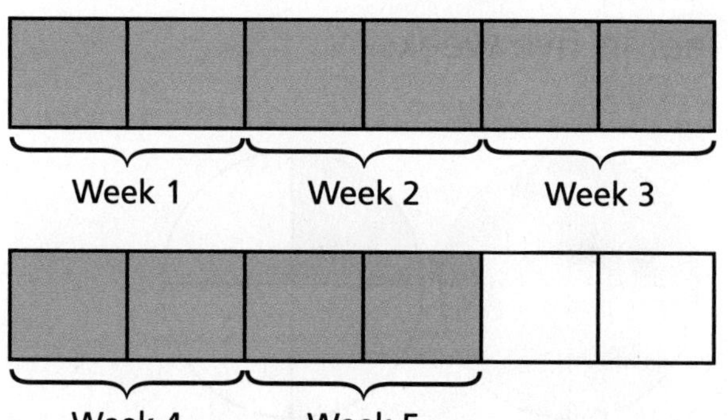

Draw a model for each problem. Then solve.

11. $4 \cdot \frac{3}{8} = ■$

12. $2 \cdot \frac{4}{5} = ■$

13. $5 \cdot \frac{2}{3} = ■$

Solve. Write your answer as a fraction.

14. $8 \cdot \frac{3}{4} = ■$

15. $18 \cdot \frac{2}{3} = ■$

16. $10 \cdot \frac{5}{6} = ■$

17. $4 \cdot \frac{5}{7} = ■$

18. $15 \cdot \frac{3}{10} = ■$

19. $7 \cdot \frac{8}{9} = ■$

Show your work on your paper or in your journal.

▶ Real World Problems

Draw a model for each problem. Then solve.

20. The five members of the Sanchez family each drank $\frac{3}{4}$ cup orange juice for breakfast. How much juice did the family drink for breakfast altogether?

21. Stella ran $\frac{1}{2}$ mile. Brian ran 7 times as far as Stella. How far did Brian run?

Write an equation. Then solve.

22. A banner has a length of 3 yards and a width of $\frac{2}{3}$ yard. What is the area of the banner?

23. The 12 members of a volleyball team had a pizza party. Each pizza was divided into 8 equal slices and each player ate 3 slices. What fraction of a pizza did the team eat altogether?

24. It took Eli's mother $\frac{1}{6}$ hour to drive him to school. It took Alex 4 times as long as this to walk to school. How long did it take Alex to walk to school?

Show your work on your paper or in your journal.

▶ Real World Problems (continued)

Write an equation. Then solve.

25. Ami has building bricks that are $\frac{5}{8}$ inch thick. She makes a stack of 15 bricks. How tall is the stack?

26. A crepe recipe calls for $\frac{3}{4}$ cups of flour. A bread recipe calls for four times this much flour. How much flour is in the bread recipe?

27. The path around a park is $\frac{7}{12}$ mile long. Nicolas ran around the park 6 times along the path. How far did he run?

▶ What's the Error?

Dear Math Students,
I have so much homework! I have assignments in math, science, and reading. I think each subject will take $\frac{1}{2}$ hour. I tried to multiply to find the total time.

$3 \cdot \frac{1}{2} = \frac{3}{6}$

That can't be right! I know $\frac{3}{6}$ is the same as $\frac{1}{2}$, so that is only $\frac{1}{2}$ hour.

What did I do wrong? How long will my homework really take?

Your friend,
Puzzled Penguin

28. Write a response to Puzzled Penguin.

▶ Multiplication Practice

Write each fraction as a sum of unit fractions and as the product of a whole number and a unit fraction.

1. $\frac{4}{7} = \blacksquare$

   $\frac{4}{7} = \blacksquare$

2. $\frac{5}{2} = \blacksquare$

   $\frac{5}{2} = \blacksquare$

3. $\frac{2}{3} = \blacksquare$

   $\frac{2}{3} = \blacksquare$

4. $\frac{6}{4} = \blacksquare$

   $\frac{6}{4} = \blacksquare$

Draw a model for each problem. Then solve.

5. $6 \cdot \frac{1}{4} = \blacksquare$

6. $6 \cdot \frac{2}{3} = \blacksquare$

7. $3 \cdot \frac{2}{9} = \blacksquare$

8. $4 \cdot \frac{4}{5} = \blacksquare$

Multiply. Write your answer as a mixed number or a whole number, when possible.

9. $20 \cdot \frac{3}{10} = \blacksquare$

10. $36 \cdot \frac{5}{9} = \blacksquare$

11. $2 \cdot \frac{2}{12} = \blacksquare$

12. $21 \cdot \frac{1}{3} = \blacksquare$

13. $16 \cdot \frac{3}{8} = \blacksquare$

14. $11 \cdot \frac{7}{10} = \blacksquare$

6-8
Class Activity

Show your work on your paper or in your journal.

► Real Word Problems

Draw a model for each problem. Then solve. Write your answer as a mixed number or a whole number, when possible.

15. Michelle has three textbooks. Each weighs $\frac{5}{8}$ pound. What is the total weight of her textbooks?

16. Mark lived in a house in the suburbs with $\frac{2}{3}$ acre of land. Then he moved to a farm in the country that had 6 times this much land. How much land is on Mark's farm?

Write an equation. Then solve. Write your answer as a mixed number or a whole number, when possible.

17. A restaurant served quiche for lunch today. Each quiche was cut into six pieces. The restaurant sold 59 pieces. How many quiches is this?

18. Zahra's dog Brutus weighed $\frac{7}{8}$ pound when he was born. Now he weighs 60 times this much. How much does Brutus weigh now?

19. Calvin made posters to advertise the school play. The posters are 2 feet long and $\frac{11}{12}$ foot wide. What is the area of each poster?

► Practice Fraction Operations

Write each fraction as a sum of fractions in two different ways.

1. $\frac{3}{10} = $ ■

2. $\frac{7}{7} = $ ■

3. $\frac{4}{5} = $ ■

4. $\frac{5}{12} = $ ■

Add or subtract.

5. $\frac{5}{8} + \frac{3}{8} = $ ■

6. $\frac{2}{10} + \frac{1}{10} = $ ■

7. $\frac{7}{9} - \frac{3}{9} = $ ■

8. $6\frac{7}{10} - 1\frac{4}{10}$

9. $5\frac{2}{3} + 4\frac{1}{3}$

10. $7\frac{1}{6} - 3\frac{2}{6}$

11. $12\frac{4}{9} + 10\frac{5}{9}$

12. $1\frac{4}{5} + 1\frac{3}{5}$

13. $7 - 1\frac{1}{4}$

Multiply. Write your answer as a mixed number or a whole number, when possible.

14. $7 \cdot \frac{1}{10} = $ ■

15. $4 \cdot \frac{2}{9} = $ ■

16. $5 \cdot \frac{3}{5} = $ ■

17. $12 \cdot \frac{3}{4} = $ ■

18. $7 \cdot \frac{5}{8} = $ ■

19. $10 \cdot \frac{5}{6} = $ ■

6-9
Class Activity

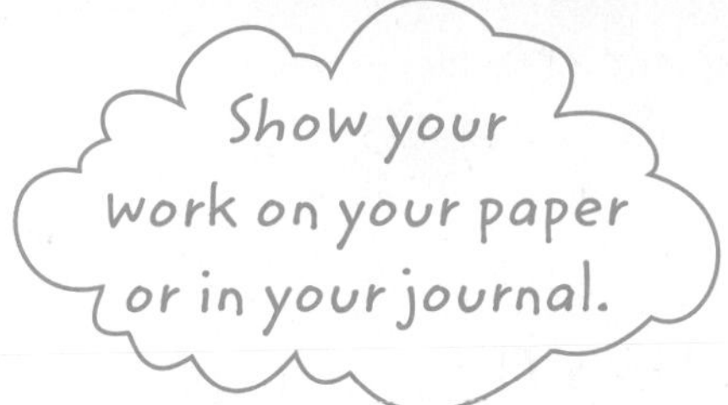
Show your
work on your paper
or in your journal.

► Real World Problems

Write an equation. Then solve.

20. Dimitri rode his bike 32 miles yesterday. He rode $12\frac{4}{5}$ miles before lunch and the rest of the distance after lunch. How far did he ride after lunch?

21. Ms. Washington is taking an accounting class. Each class is $\frac{3}{4}$ hour long. If there are 22 classes in all, how many hours will Ms. Washington spend in class?

22. Elin bought a large watermelon at the farmer's market. She cut off a $5\frac{5}{8}$-pound piece and gave it to her neighbor. She has $11\frac{5}{8}$ pounds of watermelon left. How much did the whole watermelon weigh?

23. A recipe calls for $\frac{3}{4}$ cup of whole wheat flour, $1\frac{2}{4}$ cups of white flour, and $\frac{3}{4}$ cup of rye flour. How much flour is this in all?

24. Henri spent a total of 3[illegible]/6 hours working on his science project. Kali spent $1\frac{5}{6}$ hours working on her science project. How much longer did Henri work on his project?

25. Mr. Friedman's baby daughter is $\frac{5}{9}$ yard long. Mr. Friedman's height is 4 times this much. How tall is Mr. Friedman?

26. A track is $\frac{1}{4}$ mile long. Kenny ran around the track 21 times. How far did Kenny run in all?

6-10
Class Activity

▶ Math and Vegetarian Pizza Farms

A pizza farm is a circular region of land divided into eight pie-shaped wedges or slices, such as those you would see in a pizza. There are hundreds of such farms across the United States. At a vegetarian pizza farm, each wedge or slice grows a different vegetarian ingredient used to make a pizza. Some things you might find on a vegetarian pizza farm include wheat, fruit, vegetables, Italian herbs, and dairy cows.

vegetables
wheat
Italian herbs
vegetables
vegetables
fruit
dairy cows
fruit

Write an equation to solve.

A farmer created a vegetarian pizza farm with these wedges or slices: $\frac{3}{8}$ for vegetables, $\frac{1}{8}$ for wheat, $\frac{2}{8}$ for fruit, $\frac{1}{8}$ for dairy cows, and $\frac{1}{8}$ for Italian herbs.

Show your work on your paper or in your journal.

1. What fraction of the farm is made up of fruit or vegetables?

2. What fraction of the farm is *not* made up of wheat?

3. Which wedge of the farm is bigger, the wedge for fruit or the wedge for Italian herbs? Explain.

Show your work on your paper or in your journal.

Write an equation to solve.

4. On Monday, two of the workers at the pizza farm each filled a basket with ripe tomatoes. Miles picked $15\frac{1}{6}$ pounds of tomatoes, and Anna picked $13\frac{5}{6}$ pounds of tomatoes. How many more pounds of tomatoes did Miles pick than Anna?

For Problems 5–6, use the line plot to solve.

After a field trip to a vegetarian pizza farm, Mrs. Cannon asked each of her students to use some of their study time to research different vegetarian ingredients for pizzas. The line plot below shows the amount of time each student spent researching during study time.

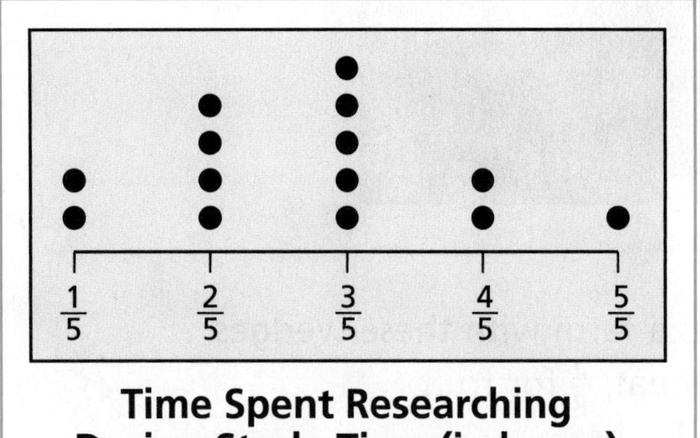

Time Spent Researching During Study Time (in hours)

5. How many students spent at least $\frac{3}{5}$ hour researching? Explain how you know.

6. How many hours in all did the students who researched for $\frac{2}{5}$ hour spend researching? Write a multiplication equation to solve.

Use the Activity Workbook Unit Test on pages 69–70.

VOCABULARY
numerator
mixed number
unit fraction

► Vocabulary

Choose the best term from the box.

1. A fraction that represents one equal part of a whole is a(n) ______■______. (Lessons 6-1)

2. A number that consists of a whole number and a fraction is a(n) ______■______. (Lesson 6-4)

► Concepts and Skills

3. Explain how to change $\frac{11}{4}$ to a mixed number. (Lesson 6-4)

4. Elias says the problem below is an addition problem. Vladmir says it is a multiplication problem. Explain why both boys are correct. (Lessons 6-3, 6-7)

   Milo practices piano $\frac{2}{3}$ hour every day. How many hours does he practice in 3 days?

Complete. (Lessons 6-1, 6-2, 6-3)

5. $\frac{3}{5} = \frac{1}{5} + \frac{1}{5} + $ ■

6. $\frac{7}{7} = \frac{2}{7} + $ ■

7. $\frac{6}{8} = \frac{4}{8} + $ ■

Write each fraction as a product of a whole number and a unit fraction. (Lessons 6-7, 6-8, 6-9)

8. $\frac{3}{8} = $ ■

9. $\frac{5}{9} = $ ■

Multiply. (Lesson 6-7, 6-8, 6-9)

10. $6 \cdot \frac{1}{5} = $ ■

11. $9 \cdot \frac{1}{3} = $ ■

12. $12 \cdot \frac{3}{4} = $ ■

13. $5 \cdot \frac{4}{7} = $ ■

Solve. (Lessons 6-3, 6-4, 6-5, 6-6)

14. $\frac{2}{5} + \frac{1}{5} = ■$

15. $\frac{7}{9} - \frac{2}{9} = ■$

16. $\frac{12}{7} + \frac{5}{7} = ■$

17. $\frac{5}{6} - \frac{4}{6} = ■$

18. $6\frac{4}{10} - 5\frac{3}{10} = ■$

19. $4\frac{3}{4} + 3\frac{1}{4} = ■$

20. $6\frac{4}{7} + 2\frac{6}{7}$

21. $5\frac{4}{9} - 1\frac{7}{9}$

22. $3 - 1\frac{2}{5}$

▶ Problem Solving

Draw a model. Then solve.

23. There is $\frac{3}{4}$ gallon of punch in a bowl. Katie added some punch. Now there is $2\frac{1}{4}$ gallons in the bowl. How much did Katie add? (Lessons 6-5, 6-6, 6-9)

24. Raul ran $\frac{4}{5}$ mile on Tuesday. He ran 4 times this far on Saturday. How far did Raul run on Saturday? (Lessons 6-7, 6-8, 6-9)

25. The line plot shows the lengths of the beads Rachel bought at the bead store today. What is the difference in length between the shortest bead and the longest bead? (Lessons 6-6)

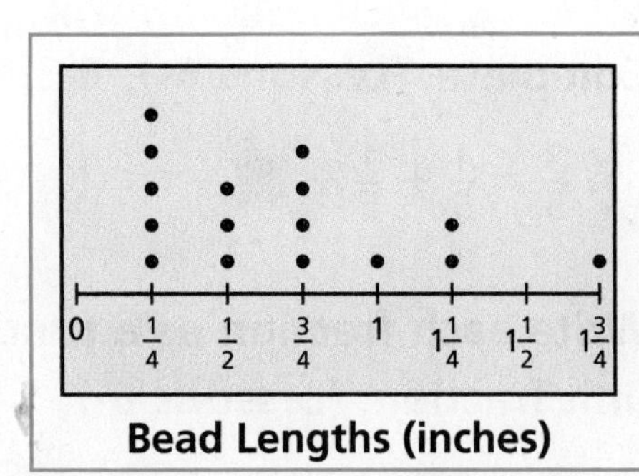

Dear Family,

In Lessons 1 through 7 of Unit 7 of *Math Expressions*, your child will build on previous experience with fractions. Your child will use both physical models and numerical methods to recognize and to find fractions equivalent to a given fraction. Your child will also compare fractions and mixed numbers, including those with like and unlike numerators and denominators.

By using fraction strips students determine how to model and compare fractions, and to find equivalent fractions. Your child will also learn how to use multiplication and division to find equivalent fractions.

Examples of Fraction Bar Modeling:

Fraction Comparisons

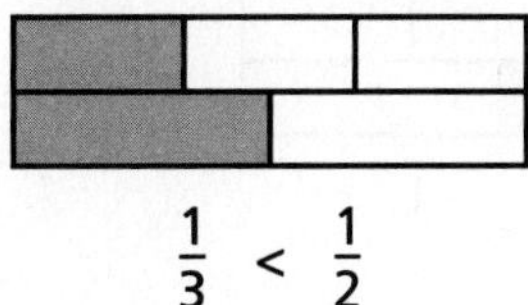

$\frac{1}{3} < \frac{1}{2}$

Equivalent Fractions

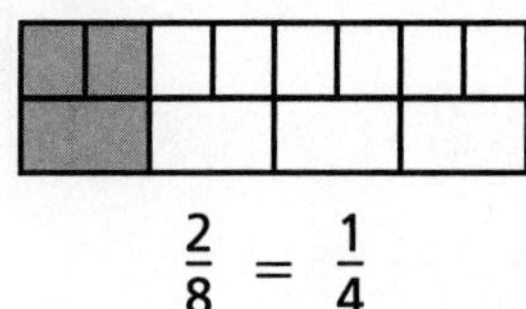

$\frac{2}{8} = \frac{1}{4}$

Share with your family the Family Letter on Activity Workbook page 71.

Your child will be introduced to the number-line model for fractions. Students name fractions corresponding to given lengths on the number line and identify lengths corresponding to given fractions. They also see that there are many equivalent fraction names for any given length.

Your child will apply this knowledge of fractions to word problems and in data displays.

If you have questions or problems, please contact me.

Thank you.

Sincerely,
Your child's teacher

COMMON CORE Lessons 1–7 of this unit include the Common Core Standards for Mathematical Content for Number and Operations—Fractions, 4.NF.1, 4.NF.2, 4.NF.5, 4.MD.4, and all Mathematical Practices.

Estimada familia:

En las lecciones 1 a 7 de la Unidad 7 de *Math Expressions*, el niño ampliará sus conocimientos previos acerca de las fracciones. Su niño usará modelos físicos y métodos numéricos para reconocer y hallar fracciones equivalentes para una fracción dada. También comparará fracciones y números mixtos, incluyendo aquellos que tengan numeradores y denominadores iguales o diferentes.

Usando tiras de fracciones, los estudiantes determinarán cómo hacer modelos y comparar fracciones y cómo hallar fracciones equivalentes. Además, aprenderán cómo usar la multiplicación y división para hallar fracciones equivalentes.

Ejemplos de modelos con barras de fracciones:

Comparar fracciones

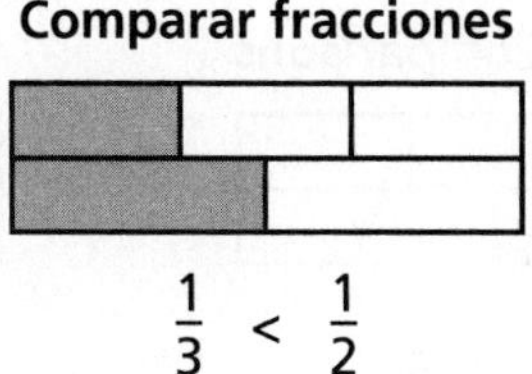

$\frac{1}{3} < \frac{1}{2}$

Fracciones equivalentes

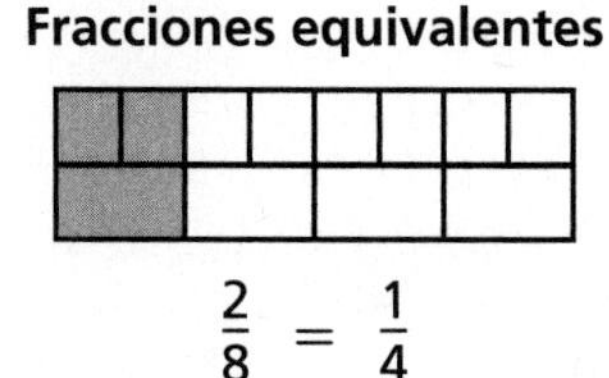

$\frac{2}{8} = \frac{1}{4}$

Su niño estudiará por primera vez el modelo de recta numérica para las fracciones. Los estudiantes nombrarán las fracciones que correspondan a determinadas longitudes en la recta numérica e identificarán longitudes que correspondan a fracciones dadas. También observarán que hay muchos nombres de fracciones equivalentes para una longitud determinada.

Su niño aplicará este conocimiento de las fracciones en problemas y en presentaciones de datos.

Si tiene alguna duda o algún comentario, por favor comuníquese conmigo.

Atentamente,
El maestro de su niño

Muestra a tu familia la Carta a la familia de la página 72 del Cuaderno de actividades y trabajo.

Las lecciones 1 a 7 de esta unidad incluyen los Common Core Standards for Mathematical Content for Number and Operations—Fractions, 4.NF.1, 4.NF.2, 4.NF.5, 4.MD.4 and all Mathematical Practices.

▶ Practice Comparing Fractions

Circle the greater fraction. Use fraction strips if you need to.

1. $\frac{1}{12}$ or $\frac{1}{2}$
2. $\frac{3}{8}$ or $\frac{1}{8}$
3. $\frac{2}{5}$ or $\frac{2}{6}$
4. $\frac{1}{3}$ or $\frac{1}{5}$
5. $\frac{4}{12}$ or $\frac{5}{12}$
6. $\frac{7}{10}$ or $\frac{5}{10}$
7. $\frac{1}{3}$ or $\frac{2}{3}$
8. $\frac{3}{6}$ or $\frac{3}{8}$

Write $>$ or $<$ to make each statement true.

9. $\frac{3}{10}$ ● $\frac{3}{8}$
10. $\frac{3}{6}$ ● $\frac{3}{5}$
11. $\frac{8}{10}$ ● $\frac{8}{12}$
12. $\frac{2}{6}$ ● $\frac{3}{6}$
13. $\frac{7}{10}$ ● $\frac{7}{8}$
14. $\frac{5}{100}$ ● $\frac{4}{100}$

▶ What's the Error?

Dear Math Students,

Yesterday, my family caught a large fish. We ate $\frac{2}{6}$ of the fish. Today, we ate $\frac{2}{4}$ of the fish. I told my mother that we ate more fish yesterday than today because 6 is greater than 4, so $\frac{2}{6}$ is greater than $\frac{2}{4}$. My mother told me I made a mistake.

Can you help me to figure out what my mistake was?

Your friend,
Puzzled Penguin

15. Write a response to Puzzled Penguin.

7-1
Class Activity

Show your work on your paper or in your journal.

▶ Make Sense of Problems

16. Explain how to compare fractions with the same denominator but different numerators.

17. Explain how to compare fractions with the same numerator but different denominators.

Solve.

18. Bao kept a list of the birds that visited his bird feeder in one day. He noticed that $\frac{2}{5}$ were finches and $\frac{2}{6}$ were wrens. Did more finches or more wrens visit the bird feeder? Tell how you know.

19. Mariel had a box of baseball cards. She kept $\frac{3}{8}$ of the cards and gave $\frac{5}{8}$ of the cards to Javier. Who had more of the cards? Explain.

20. Write the fractions $\frac{10}{12}$, $\frac{5}{12}$, and $\frac{7}{12}$ in order from least to greatest.

▶ Discuss Number Lines

The number line below shows the fourths between 0 and 1. Discuss how the number line is like and unlike the fraction bar above it.

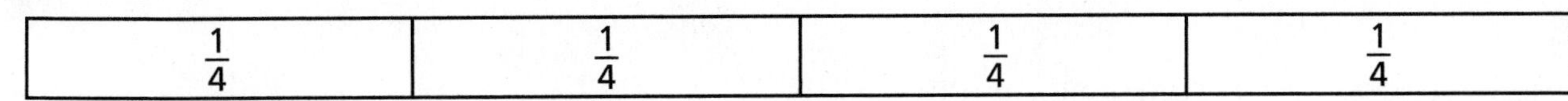

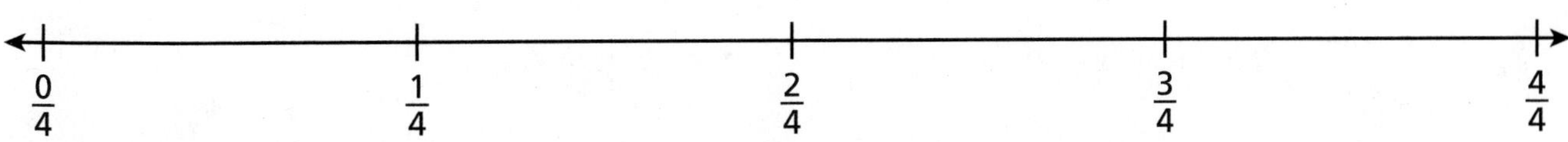

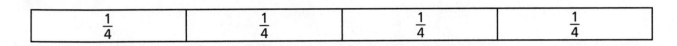

These number lines are divided to show different fractions.

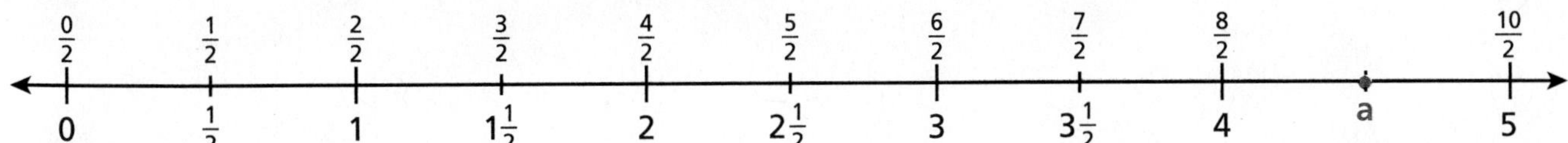

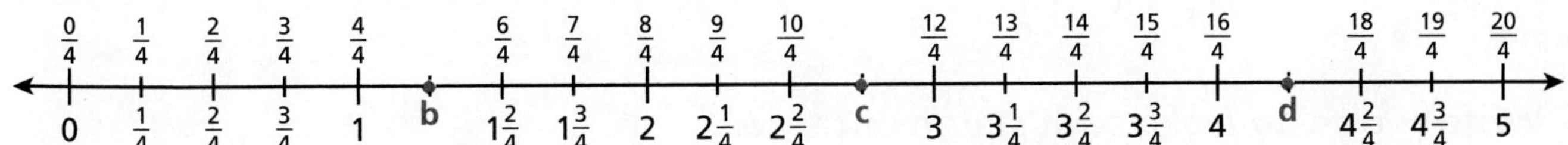

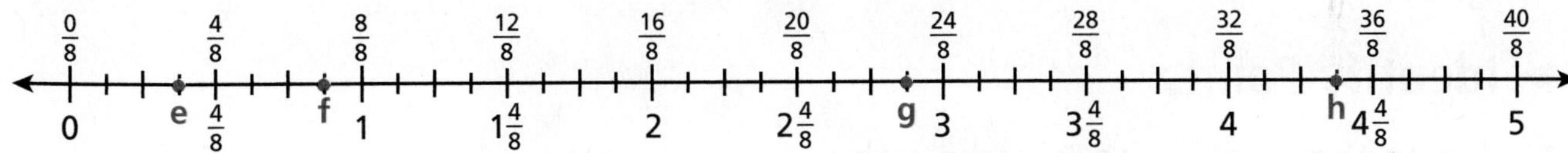

Write > or < to make each statement true.

1. $\frac{3}{4}$ ● $\frac{5}{2}$
2. $\frac{15}{4}$ ● $\frac{20}{8}$
3. $\frac{10}{4}$ ● $\frac{24}{8}$
4. $2\frac{4}{8}$ ● $1\frac{3}{4}$

▶ Identify Points

5. Write the fraction or mixed number for each lettered point above.

a. ■ b. ■ c. ■ d. ■

e. ■ f. ■ g. ■ h. ■

Use Activity Workbook page 75.

▶ Number Lines for Thirds and Sixths

Tell how many equal parts are between zero and 1. Then write fraction labels above the equal parts.

6.

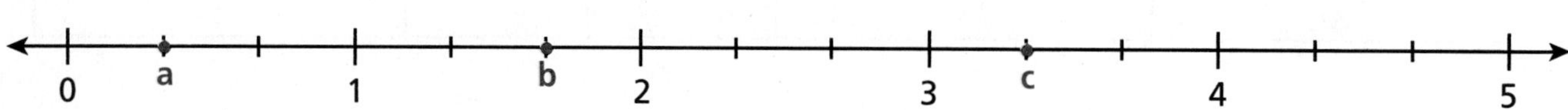

7.

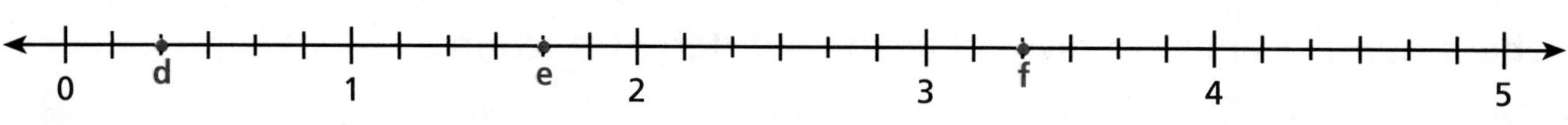

8.

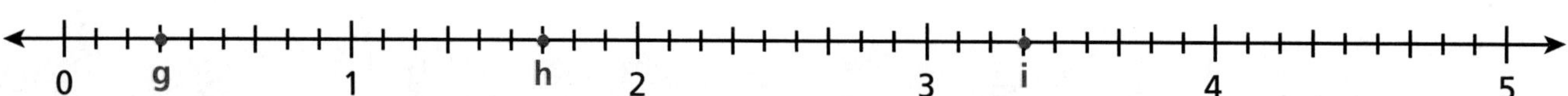

Write > or < to make each statement true.

9. $\frac{4}{3}$ ● $\frac{7}{6}$

10. $\frac{8}{3}$ ● $\frac{18}{6}$

11. $3\frac{5}{6}$ ● $3\frac{2}{3}$

▶ Identify Points

12. Write the fraction or mixed number for each lettered point above. Describe any patterns you see with the class.

a. ■ b. ■ c. ■

d. ■ e. ■ f. ■

■ h. ■ i. ■

Mark and label the letter of each fraction or mixed number on the number line.

13. 0 1 2 3 4 5 6

a. $\frac{1}{5}$ b. $\frac{7}{10}$ c. $1\frac{2}{5}$ d. $2\frac{1}{2}$

e. $3\frac{3}{10}$ f. $4\frac{2}{5}$ g. $4\frac{9}{10}$ h. $5\frac{1}{2}$

► Fractions and Benchmarks

Decide if each fraction is closer to 0 or closer to 1. Write *closer to 0* or *closer to 1*.

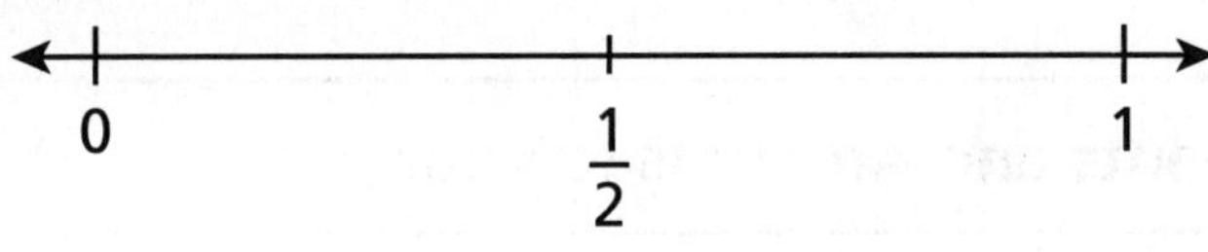

14. $\frac{1}{4}$

15. $\frac{3}{4}$

16. $\frac{7}{8}$

Write > or < to make each statement true.

17. $\frac{5}{8}$ ● $\frac{11}{12}$

18. $\frac{7}{12}$ ● $\frac{1}{8}$

19. $\frac{3}{8}$ ● $\frac{1}{6}$

► What's the Error?

Dear Math Students,

I am baking cookies. My recipe calls for $\frac{5}{8}$ cup of walnuts. Walnuts come in $\frac{1}{2}$-pound bags and 1-pound bags. My friend says that $\frac{5}{8}$ is closer to $\frac{1}{2}$ than it is to 1, so I should buy a $\frac{1}{2}$-pound bag. I think my friend is wrong.

Do you agree with me or with my friend? Can you help me decide what size bag of walnuts I should buy?

Your friend,
Puzzled Penguin

20. Write a response to Puzzled Penguin.

▶ Use Benchmarks

The list below shows a variety of cooking ingredients and amounts.

Ingredients and Amounts (c = cup)	
wheat flour–$1\frac{5}{8}$ c	white flour–$\frac{5}{6}$ c
sugar–$1\frac{1}{8}$ c	cornstarch–$\frac{3}{8}$ c
oat bran–$1\frac{4}{5}$ c	water–$\frac{2}{6}$ c

Decide if each amount is closer to $\frac{1}{2}$ cup, $1\frac{1}{2}$ cups, or 2 cups. Write *closer to* $\frac{1}{2}$ *c, closer to* $1\frac{1}{2}$ *c,* or *closer to 2 c.*

21. wheat flour
22. white flour
23. sugar
24. cornstarch
25. oat bran
26. water

Decide which ingredient represents a greater amount.

27. sugar or water
28. sugar or wheat flour
29. cornstarch or sugar
30. wheat flour or white flour
31. sugar or oat bran
32. oat bran or white flour

Show your work on your paper or in your journal.

▶ Compare Fractions of Different-Size Wholes

Jon and his five friends want sandwiches. They make two sandwiches: one on a short loaf of bread and one on a longer loaf. Jon cuts each sandwich into 6 pieces. His friends think the pieces are not the same size.

1. Are Jon's friends correct? Explain.

2. What can Jon do to make sure everyone gets the same amount of food?

Hattie's dad orders one small, one medium, and one large pizza. He divides each pizza into 8 equal pieces. Hattie takes $\frac{1}{8}$ of the small pizza and her friend takes $\frac{1}{8}$ of the large pizza.

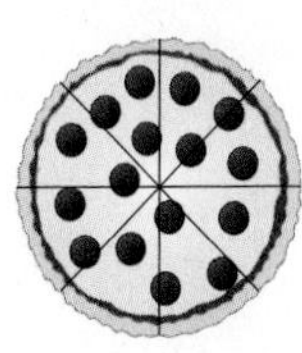

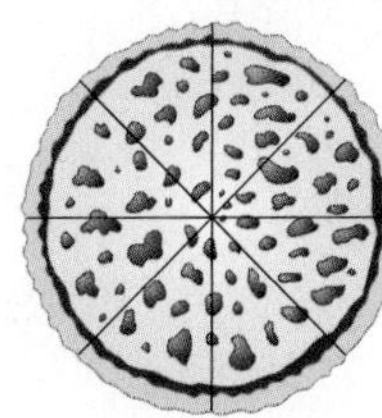

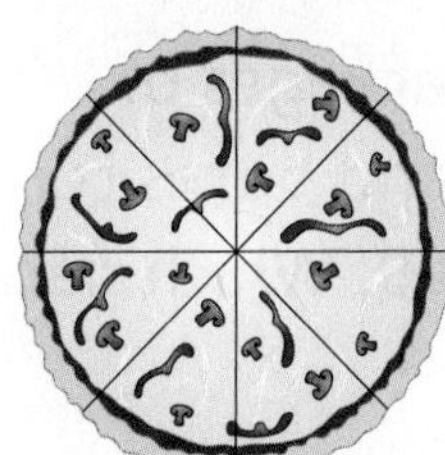

3. Hattie says she has less pizza than her friend. Is she correct? Explain.

4. What do these problems tell us about fractions?

7-3
Class Activity

Show your
work on your paper
or in your journal.

► Fraction Word Problems

Solve.

5. A shelter had 4 spaniel puppies and 6 beagle puppies. Jack adopted $\frac{1}{2}$ of the spaniel puppies, and Carmen adopted $\frac{1}{2}$ of the beagle puppies. Who adopted more puppies? How do you know?

6. Julio planted 16 daisies and 10 sunflowers. His neighbor's goat ate 5 daisies and 5 sunflowers. Did the goat eat a greater fractional part of the daisies or the sunflowers? Explain.

7. A fruit market sells two different packages of oranges. Bags contain 12 oranges, and boxes contain 15 oranges. Both packages cost $3.00. Which package is a better buy? Why?

8. The fourth grade has three running teams. Each team has 12 runners. In a race, $\frac{1}{4}$ of Team A, $\frac{1}{3}$ of Team B, and $\frac{1}{6}$ of Team C passed the first water stop at the same time. Which team had the most runners at the first water stop at that time? Explain.

VOCABULARY
equivalent fractions

► Equivalent Fractions

Read and discuss the problem situation.

Luis works summers at Maria's Fruit Farm. One day, Maria agreed to give Luis extra pay if he could sell $\frac{2}{3}$ of her supply of peaches. They started with 12 bags of peaches, and Luis sold 8 of them.

1. Luis said to Maria, "Eight bags is $\frac{8}{12}$ of the 12 bags you wanted to sell. I think $\frac{2}{3}$ is the same as $\frac{8}{12}$. I can show you why." Luis made this drawing. Did Luis earn his pay?

$\frac{1}{3}$				$\frac{1}{3}$				$\frac{1}{3}$			
$\frac{1}{12}$	$\frac{1}{12}$	$\frac{1}{12}$	$\frac{1}{12}$	$\frac{1}{12}$	$\frac{1}{12}$	$\frac{1}{12}$	$\frac{1}{12}$	$\frac{1}{12}$	$\frac{1}{12}$	$\frac{1}{12}$	$\frac{1}{12}$

$$\frac{2}{3} = \frac{8}{12}$$

Two fractions that represent the same part of a whole are **equivalent fractions**. The fractions $\frac{2}{3}$ and $\frac{8}{12}$ are equivalent.

2. Maria said, "You are just fracturing each third into 4 twelfths. You can show what you did using numbers." Here's what Maria wrote:

$$\frac{2}{3} = \frac{2 \times 4}{3 \times 4} = \frac{8}{12}$$

Discuss what Maria did. How does multiplying the numerator and denominator by 4 affect the fraction?

▶ Use Fraction Bars to Find Equivalent Fractions

3. How do these fraction bars show equivalent fractions for $\frac{1}{3}$?

$\frac{1}{3}$	$\frac{1}{3}$	$\frac{1}{3}$

$\frac{1}{6}$	**$\frac{1}{6}$**	$\frac{1}{6}$	$\frac{1}{6}$	$\frac{1}{6}$	$\frac{1}{6}$

$\frac{1}{9}$	**$\frac{1}{9}$**	**$\frac{1}{9}$**	$\frac{1}{9}$	$\frac{1}{9}$	$\frac{1}{9}$	$\frac{1}{9}$	$\frac{1}{9}$	$\frac{1}{9}$

$\frac{1}{12}$	**$\frac{1}{12}$**	**$\frac{1}{12}$**	**$\frac{1}{12}$**	$\frac{1}{12}$	$\frac{1}{12}$	$\frac{1}{12}$	$\frac{1}{12}$	$\frac{1}{12}$	$\frac{1}{12}$	$\frac{1}{12}$	$\frac{1}{12}$

$\frac{1}{15}$	**$\frac{1}{15}$**	**$\frac{1}{15}$**	**$\frac{1}{15}$**	**$\frac{1}{15}$**	$\frac{1}{15}$	$\frac{1}{15}$	$\frac{1}{15}$	$\frac{1}{15}$	$\frac{1}{15}$	$\frac{1}{15}$	$\frac{1}{15}$	$\frac{1}{15}$	$\frac{1}{15}$	$\frac{1}{15}$

$\frac{1}{18}$	**$\frac{1}{18}$**	**$\frac{1}{18}$**	**$\frac{1}{18}$**	**$\frac{1}{18}$**	**$\frac{1}{18}$**	$\frac{1}{18}$	$\frac{1}{18}$	$\frac{1}{18}$	$\frac{1}{18}$	$\frac{1}{18}$	$\frac{1}{18}$	$\frac{1}{18}$	$\frac{1}{18}$	$\frac{1}{18}$	$\frac{1}{18}$	$\frac{1}{18}$	$\frac{1}{18}$

4. You can show how to find fractions equivalent to $\frac{1}{3}$ numerically. Fill in the blanks and finish the equations. Then explain how these fraction equations show equivalent fractions.

2 equal parts	3 equal parts	4 equal parts	$\square$ equal parts	$\square$ equal parts
$\times 2$	$\times 3$	$\times \square$	$\times \square$	$\times \square$
$\frac{1 \times 2}{3 \times 2} = \frac{2}{6}$	$\frac{1 \times \square}{3 \times \square} = \frac{\square}{9}$	$\frac{1 \times \square}{3 \times \square} = \frac{\square}{12}$	$\frac{1 \times \square}{3 \times \square} = \frac{\square}{15}$	$\frac{1 \times \square}{3 \times \square} = \frac{\square}{18}$

5. Tell whether the fractions are equivalent.

a. $\frac{1}{6}$ and $\frac{2}{12}$

b. $\frac{3}{6}$ and $\frac{5}{9}$

c. $\frac{6}{12}$ and $\frac{8}{15}$

▸ Use a Multiplication Table to Find Equivalent Fractions

The table on the right shows part of the multiplication table at the left. You can find fractions equivalent to $\frac{1}{3}$ by using the products in the rows for the factors 1 and 3.

×	1	2	3	4	5	6	7	8	9	10
1	1	2	3	4	5	6	7	8	9	10
2	2	4	6	8	10	12	14	16	18	20
3	3	6	9	12	15	18	21	24	27	30
4	4	8	12	16	20	24	28	32	36	40
5	5	10	15	20	25	30	35	40	45	50
6	6	12	18	24	30	36	42	48	54	60
7	7	14	21	28	35	42	49	56	63	70
8	8	16	24	32	40	48	56	64	72	80
9	9	18	27	36	45	54	63	72	81	90
10	10	20	30	40	50	60	70	80	90	100

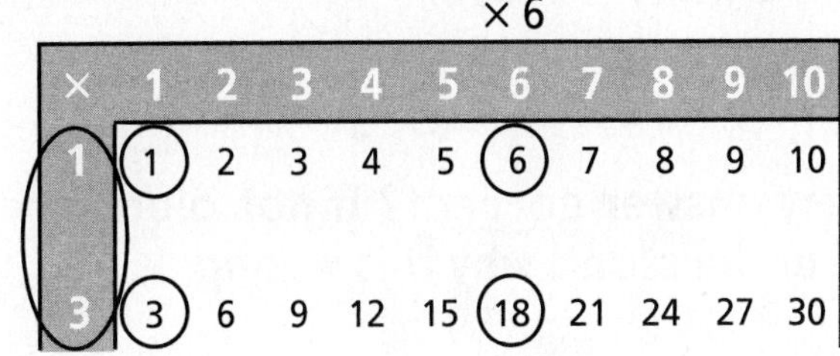
× 6

×	1	2	3	4	5	6	7	8	9	10
1	1	2	3	4	5	6	7	8	9	10
3	3	6	9	12	15	18	21	24	27	30

$$\frac{1 \times 6}{3 \times 6} = \frac{6}{18}$$

$$\frac{6 \div 6}{18 \div 6} = \frac{1}{3}$$

$\frac{1}{3}$	$\frac{1}{3}$	$\frac{1}{3}$

$\frac{1}{18}$	$\frac{1}{18}$	$\frac{1}{18}$	$\frac{1}{18}$	$\frac{1}{18}$	$\frac{1}{18}$	$\frac{1}{18}$	$\frac{1}{18}$	$\frac{1}{18}$	$\frac{1}{18}$	$\frac{1}{18}$	$\frac{1}{18}$	$\frac{1}{18}$	$\frac{1}{18}$	$\frac{1}{18}$	$\frac{1}{18}$	$\frac{1}{18}$	$\frac{1}{18}$

Complete each fraction equation. Look in the top row of the table above to find the multiplier.

6. $\frac{1 \times \square}{3 \times \square} = \frac{4}{12}$

7. $\frac{1 \times \square}{3 \times \square} = \frac{9}{27}$

8. $\frac{1 \times \square}{3 \times \square} = \frac{2}{6}$

9. $\frac{1 \times \square}{4 \times \square} = \frac{3}{12}$

10. $\frac{3 \times \square}{10 \times \square} = \frac{30}{100}$

11. $\frac{5 \times \square}{8 \times \square} = \frac{30}{48}$

12. Tell whether the fractions are equivalent.

    a. $\frac{3}{4}$ and $\frac{12}{16}$

    b. $\frac{1}{2}$ and $\frac{5}{12}$

    c. $\frac{9}{10}$ and $\frac{90}{100}$

▶ What's the Error?

Dear Students,

I tried to find a fraction equivalent to $\frac{2}{3}$. Here's what I wrote.

$$\frac{2}{3} = \frac{5}{6}$$

Is my answer correct? If not, please help me understand why it is wrong.

Thank you.
Puzzled Penguin

13. Write a response to Puzzled Penguin.

▶ Practice

Find a fraction equivalent to the given fraction.

14. $\frac{1}{4}$ $\frac{1 \times 2}{4 \times 2} = \frac{2}{\blacksquare}$

15. $\frac{3}{8}$ $\frac{3 \times 3}{8 \times 3} = \frac{9}{\blacksquare}$

16. $\frac{3}{10}$

17. $\frac{3}{4}$

18. $\frac{4}{5}$

19. $\frac{7}{12}$

20. $\frac{5}{6}$

21. $\frac{7}{8}$

22. Write two fractions with the denominator 100: one equivalent to $\frac{1}{4}$ and one equivalent to $\frac{7}{10}$.

7-5
Class Activity

Use Activity Workbook page 76.

VOCABULARY
simplify

▶ Simplify Fractions

Simplifying a fraction means finding an equivalent fraction with a lesser numerator and denominator. Simplifying a fraction results in an equivalent fraction with fewer but greater unit fractions.

1. Maria had 12 boxes of apricots. She sold 10 of the boxes. Write the fraction of the boxes sold, and lightly shade the twelfths fraction bar to show this fraction.

   Fraction sold: $\square$

2. Group the twelfths to form an equivalent fraction with a lesser denominator. Show the new fraction by dividing, labeling, and lightly shading the blank fraction bar.

   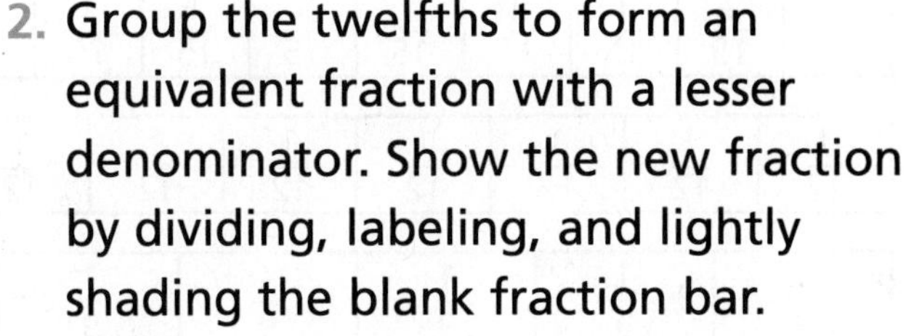

   Fraction sold: $\square$

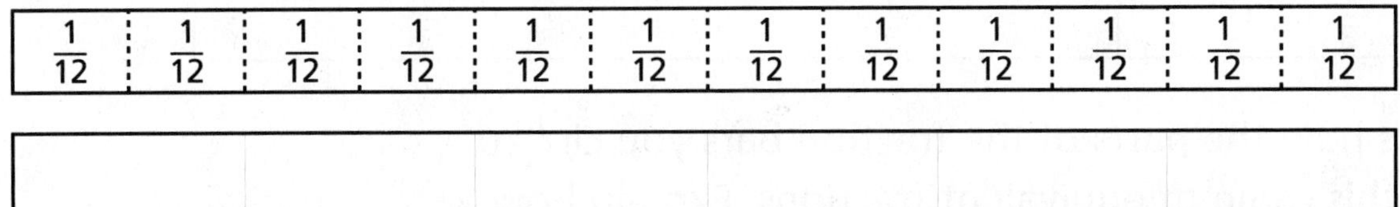

3. In Problem 2, you formed groups of twelfths to get a greater unit fraction. How many twelfths are in each group? In other words, what is the *group size*?

4. Show how you can find the equivalent fraction by dividing the numerator and denominator by the group size.

   $\frac{10}{12} = \frac{10 \div \square}{12 \div \square} = \frac{\square}{\square}$

Use what you know to find these equivalent fractions. You may want to sketch a thirds fraction bar below the two fraction bars above.

5. $\frac{8}{12} = \frac{\square}{6} = \frac{\square}{3}$

6. $\frac{4}{12} = \frac{\square}{6} = \frac{\square}{3}$

7. $\frac{20}{12} = \frac{\square}{6} = \frac{\square}{3} = \square\frac{\square}{3}$

Use Activity Workbook page 77.

▶ Use Fraction Bars to Find Equivalent Fractions

8. Look at the thirds bar. Circle enough unit fractions on each of the other bars to equal $\frac{1}{3}$.

$\frac{1}{18}$	$\frac{1}{18}$	$\frac{1}{18}$	$\frac{1}{18}$	$\frac{1}{18}$	$\frac{1}{18}$	$\frac{1}{18}$	$\frac{1}{18}$	$\frac{1}{18}$	$\frac{1}{18}$	$\frac{1}{18}$	$\frac{1}{18}$	$\frac{1}{18}$	$\frac{1}{18}$	$\frac{1}{18}$	$\frac{1}{18}$	$\frac{1}{18}$	$\frac{1}{18}$

$\frac{1}{15}$	$\frac{1}{15}$	$\frac{1}{15}$	$\frac{1}{15}$	$\frac{1}{15}$	$\frac{1}{15}$	$\frac{1}{15}$	$\frac{1}{15}$	$\frac{1}{15}$	$\frac{1}{15}$	$\frac{1}{15}$	$\frac{1}{15}$	$\frac{1}{15}$	$\frac{1}{15}$	$\frac{1}{15}$

$\frac{1}{12}$	$\frac{1}{12}$	$\frac{1}{12}$	$\frac{1}{12}$	$\frac{1}{12}$	$\frac{1}{12}$	$\frac{1}{12}$	$\frac{1}{12}$	$\frac{1}{12}$	$\frac{1}{12}$	$\frac{1}{12}$	$\frac{1}{12}$

$\frac{1}{9}$	$\frac{1}{9}$	$\frac{1}{9}$	$\frac{1}{9}$	$\frac{1}{9}$	$\frac{1}{9}$	$\frac{1}{9}$	$\frac{1}{9}$	$\frac{1}{9}$

$\frac{1}{6}$	$\frac{1}{6}$	$\frac{1}{6}$	$\frac{1}{6}$	$\frac{1}{6}$	$\frac{1}{6}$

$\frac{1}{3}$	$\frac{1}{3}$	$\frac{1}{3}$

9. Discuss how the parts of the fraction bars you circled show this chain of equivalent fractions. Explain how each different group of unit fractions is equal to $\frac{1}{3}$.

$$\frac{6}{18} = \frac{5}{15} = \frac{4}{12} = \frac{3}{9} = \frac{2}{6} = \frac{1}{3}$$

10. Write the group size for each fraction in the chain of equivalent fractions. The first one is done for you.

6

11. Complete each equation by showing how you use group size to simplify. The first one is done for you.

$\frac{6 \div 6}{18 \div 6} = \frac{1}{3}$ $\qquad$ $\frac{5 \div \square}{15 \div \square} = \frac{1}{3}$ $\qquad$ $\frac{4 \div \square}{12 \div \square} = \frac{1}{3}$

$\frac{3 \div \square}{9 \div \square} = \frac{1}{3}$ $\qquad$ $\frac{2 \div \square}{6 \div \square} = \frac{1}{3}$

► Use a Multiplication Table to Find Equivalent Fractions

Multiplication table rows show relationships among equivalent fractions.

12. What happens to the fractions as you move from right to left? How does the size of the unit fraction change? How does the number of unit fractions change?

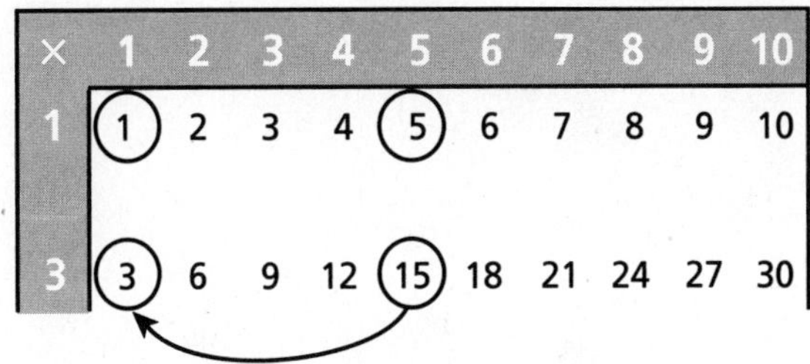

×	1	2	3	4	5	6	7	8	9	10
1	1	2	3	4	5	6	7	8	9	10
3	3	6	9	12	15	18	21	24	27	30

Simplify fractions.

$$\frac{5}{15} = \frac{5 \div 5}{15 \div 5} = \frac{1}{3}$$

13. What happens to the fractions as you move from left to right? How does the size of the unit fraction change? How does the number of unit fractions change?

×	1	2	3	4	5	6	7	8	9	10
1	1	2	3	4	5	6	7	8	9	10
3	3	6	9	12	15	18	21	24	27	30

Unsimplify fractions.

$$\frac{1}{3} = \frac{1 \times 5}{3 \times 5} = \frac{5}{15}$$

► Use a Multiplication Table to Find Equivalent Fractions (continued)

Here are two more rows from the multiplication table moved together. These rows can be used to generate a chain of fractions equivalent to $\frac{4}{8}$.

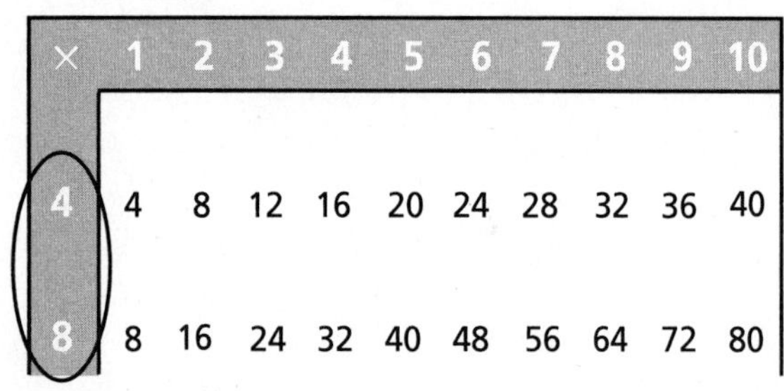

×	1	2	3	4	5	6	7	8	9	10
4	4	8	12	16	20	24	28	32	36	40
8	8	16	24	32	40	48	56	64	72	80

Complete each equation.

14. $\frac{4 \times \square}{8 \times \square} = \square$

15. $\frac{4 \times \square}{8 \times \square} = \square$

16. $\frac{20 \div \square}{40 \div \square} = \square$

17. $\frac{36 \div \square}{72 \div \square} = \square$

18. $\frac{12 \div \square}{24 \div \square} = \square$

19. $\frac{24 \div \square}{48 \div \square} = \square$

Complete each chain of equivalent fractions.

20. $\frac{1}{2} = \frac{\square}{4} = \frac{\square}{8}$

21. $\frac{\square}{5} = \frac{\square}{10} = \frac{\square}{20}$

22. $\frac{\square}{3} = \frac{\square}{6} = \frac{\square}{12}$

23. $\frac{\square}{2} = \frac{\square}{4} = \frac{\square}{8}$

► Practice Simplifying Fractions

Simplify each fraction.

24. $\frac{8 \div \square}{10 \div \square} = \square$

25. $\frac{6 \div \square}{8 \div \square} = \square$

26. $\frac{15 \div \square}{40 \div \square} = \square$

27. $\frac{10 \div \square}{12 \div \square} = \square$

28. $\frac{8 \div \square}{12 \div \square} = \square$

29. $\frac{20 \div \square}{30 \div \square} = \square$

30. $\frac{40 \div \square}{100 \div \square} = \square$

31. $\frac{75 \div \square}{100 \div \square} = \square$

Complete each chain of equivalent fractions.

32. $\frac{16}{8} = \frac{\square}{4} = \frac{\square}{2}$

33. $\frac{\square}{20} = \frac{\square}{10} = \frac{\square}{5}$

34. $\frac{\square}{12} = \frac{\square}{6} = \frac{\square}{3}$

35. $\frac{\square}{8} = \frac{\square}{4} = \frac{\square}{2}$

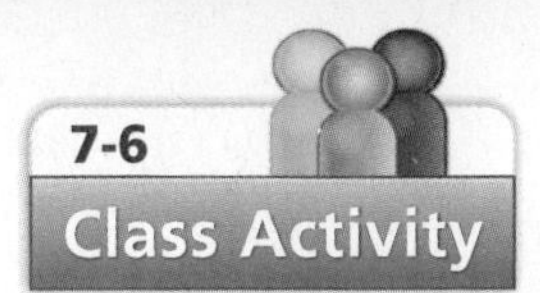

▶ Compare Fractions Using Fraction Strips and Number Lines

1. Use the number lines to compare the fractions $\frac{4}{5}$ and $\frac{7}{10}$.

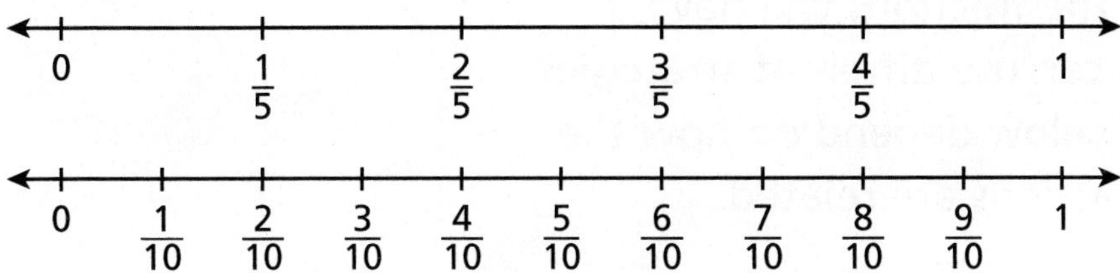

2. Use the fraction strips to compare the fractions $\frac{3}{4}$ and $\frac{5}{6}$.

$\frac{1}{4}$	$\frac{1}{4}$	$\frac{1}{4}$	$\frac{1}{4}$

$\frac{1}{6}$	$\frac{1}{6}$	$\frac{1}{6}$	$\frac{1}{6}$	$\frac{1}{6}$	$\frac{1}{6}$

$\frac{1}{12}$	$\frac{1}{12}$	$\frac{1}{12}$	$\frac{1}{12}$	$\frac{1}{12}$	$\frac{1}{12}$	$\frac{1}{12}$	$\frac{1}{12}$	$\frac{1}{12}$	$\frac{1}{12}$	$\frac{1}{12}$	$\frac{1}{12}$

Compare. Write $>$, $<$, or $=$.

3. $\frac{3}{4}$ ● $\frac{7}{12}$
4. $\frac{3}{5}$ ● $\frac{7}{12}$
5. $\frac{3}{5}$ ● $\frac{6}{10}$
6. $\frac{2}{5}$ ● $\frac{3}{6}$
7. $\frac{4}{10}$ ● $\frac{1}{5}$
8. $\frac{2}{10}$ ● $\frac{3}{8}$

VOCABULARY
common denominator

▶ Compare Fractions Using Common Denominators

You can compare two fractions with different denominators by writing equivalent fractions that use the same unit fraction. The fractions will have a **common denominator**. You can use different strategies to do this. The ones shown below depend on how the denominators of the two fractions are related.

Case 1: One denominator is a factor of the other. **Possible Strategy:** Use the greater denominator as the common denominator.	**Example** Compare $\frac{3}{5}$ and $\frac{5}{10}$. Use 10 as the common denominator. $\frac{3 \times 2}{5 \times 2} = \frac{6}{10}$ $\frac{6}{10} > \frac{5}{10}$, so $\frac{3}{5} > \frac{5}{10}$.
Case 2: The only number that is a factor of both denominators is 1. **Possible Strategy:** Use the product of the denominators as the common denominator.	**Example** Compare $\frac{5}{8}$ and $\frac{4}{5}$. Use 5×8, or 40, as the common denominator. $\frac{5 \times 5}{8 \times 5} = \frac{25}{40}$ $\frac{4 \times 8}{5 \times 8} = \frac{32}{40}$ $\frac{25}{40} < \frac{32}{40}$, so $\frac{5}{8} < \frac{4}{5}$.
Case 3: There is a number besides 1 that is a factor of both denominators. **Possible Strategy:** Use a common denominator that is less than the product of the denominators.	**Example** Compare $\frac{5}{8}$ and $\frac{7}{12}$. 24 is a common multiple of 8 and 12. Use 24 as the common denominator. $\frac{5 \times 3}{8 \times 3} = \frac{15}{24}$ $\frac{7 \times 2}{12 \times 2} = \frac{14}{24}$ $\frac{15}{24} > \frac{14}{24}$, so $\frac{5}{8} > \frac{7}{12}$.

Compare. Write >, <, or =.

9. $\frac{3}{5}$ ● $\frac{2}{3}$

10. $\frac{10}{12}$ ● $\frac{5}{6}$

11. $\frac{3}{4}$ ● $\frac{8}{10}$

12. $\frac{4}{5}$ ● $\frac{75}{100}$

13. $\frac{5}{8}$ ● $\frac{3}{5}$

14. $\frac{2}{3}$ ● $\frac{7}{10}$

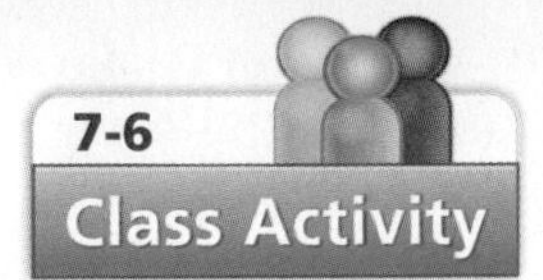

► What's the Error?

Dear Math Students,

My brother had a bowl of cherries to share. My brother ate $\frac{3}{8}$ of the cherries. I ate $\frac{2}{5}$ of the cherries. I wrote two fractions with a common denominator and compared them.

$\frac{3}{8 \times 5} = \frac{3}{40}$ and $\frac{2}{5 \times 8} = \frac{2}{40}$

$\frac{3}{40} > \frac{2}{40}$, so $\frac{3}{8} > \frac{2}{5}$.

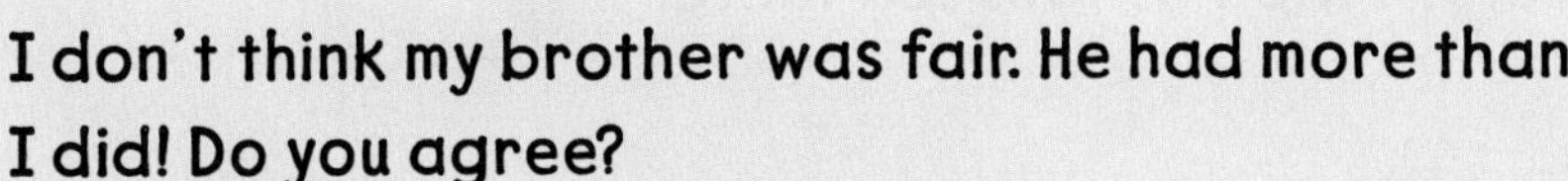
I don't think my brother was fair. He had more than I did! Do you agree?

Your friend,
Puzzled Penguin

15. Write a response to Puzzled Penguin.

► Practice

Compare.

16. $\frac{3}{6}$ ● $\frac{5}{10}$

17. $\frac{10}{12}$ ● $\frac{7}{8}$

18. $\frac{2}{6}$ ● $\frac{1}{5}$

19. $\frac{3}{8}$ ● $\frac{1}{4}$

20. $\frac{3}{10}$ ● $\frac{25}{100}$

21. $\frac{6}{12}$ ● $\frac{2}{3}$

22. $\frac{2}{5}$ ● $\frac{35}{100}$

23. $\frac{5}{12}$ ● $\frac{9}{10}$

24. $\frac{45}{100}$ ● $\frac{5}{10}$

25. $\frac{4}{5}$ ● $\frac{11}{12}$

26. $\frac{3}{12}$ ● $\frac{6}{8}$

27. $\frac{11}{12}$ ● $\frac{9}{10}$

Show your work on your paper or in your journal.

► Practice (continued)

Solve.

28. Alexi and Kirsti are painting a fence around their garden. Alexi has painted $\frac{3}{8}$ of the fence. Kirsti has painted $\frac{5}{12}$ of the fence. Who has painted more of the fence?

29. Esther and Lavinia have the same math homework. Esther has finished $\frac{7}{8}$ of the homework. Lavinia has finished $\frac{3}{5}$ of the homework. Who has finished more of the homework?

30. Avram and Anton live on the same street. Avram's house is $\frac{3}{4}$ mile from the school. Anton's house is $\frac{7}{10}$ mile from the school. Which boy's house is a greater distance from the school?

31. Leola is reading a book. On Friday, she read $\frac{25}{100}$ of the book. On Saturday, she read $\frac{3}{8}$ of the book. On which day did she read more of the book?

► Adding Fractions

Add.

32. $\frac{2}{10} + \frac{3}{100} = \blacksquare$

33. $\frac{17}{100} + \frac{7}{10} = \blacksquare$

34. $\frac{9}{10} + \frac{33}{100} = \blacksquare$

► Use Line Plots to Solve Problems

A line plot is a graph that shows data using a number line. Mateo wants to bake raisin bread. He has several recipes that each make one loaf of bread. The line plot shows the numbers of cups of sugar in the recipes.

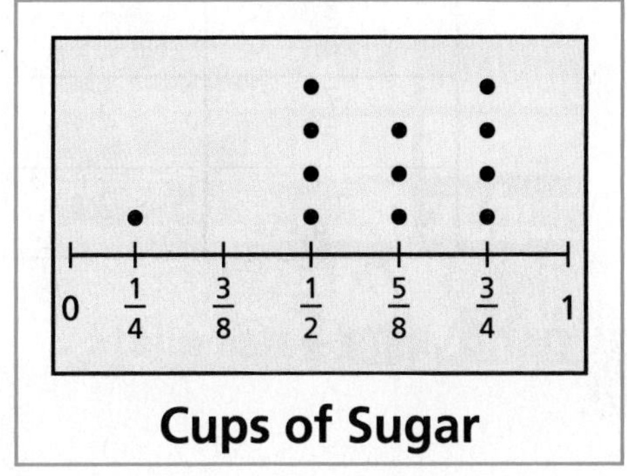

Cups of Sugar

1. How many recipes for raisin bread does Mateo have?

2. How many of the recipes have more than $\frac{1}{2}$ cup of sugar?

3. What is the least amount of sugar in any recipe?

4. How much less sugar is in a recipe with the least sugar than in a recipe with the most sugar?

5. Mateo wants to try all the recipes with exactly $\frac{5}{8}$ cup of sugar. How much sugar does he need?

6. How much sugar would you expect any recipe for raisin bread to need? Explain your thinking.

7-7
Class Activity

Use Activity Workbook page 78.

► Make a Line Plot

Mai cut up strips of color paper to make a collage. The lengths of the unused pieces are shown in the table.

Length (in inches)	Number of Pieces
$\frac{1}{2}$	4
$\frac{5}{8}$	2
$\frac{3}{4}$	2
$\frac{7}{8}$	3
$1\frac{1}{4}$	2

7. Make a line plot to display the data.

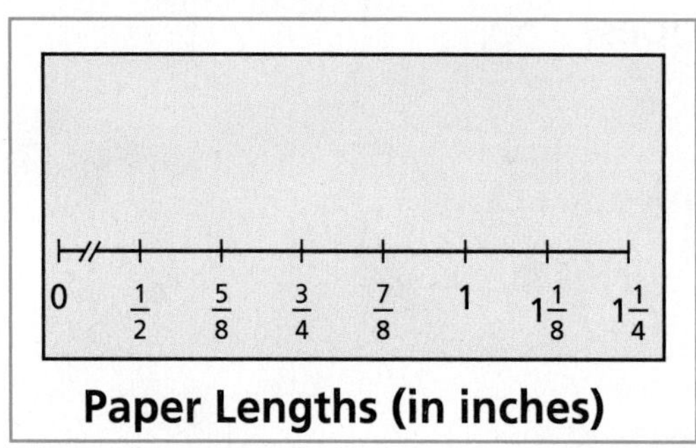

Paper Lengths (in inches)

8. Mai placed the shortest pieces in a row end to end. How long was the row?

A group of students measured the widths of their hands. The measurements are shown in the table.

Width (in inches)	Number of Students
$2\frac{1}{4}$	1
$2\frac{3}{8}$	2
$2\frac{1}{2}$	2
$2\frac{5}{8}$	4
$2\frac{3}{4}$	2
$2\frac{7}{8}$	1

9. Make a line plot to display the data.

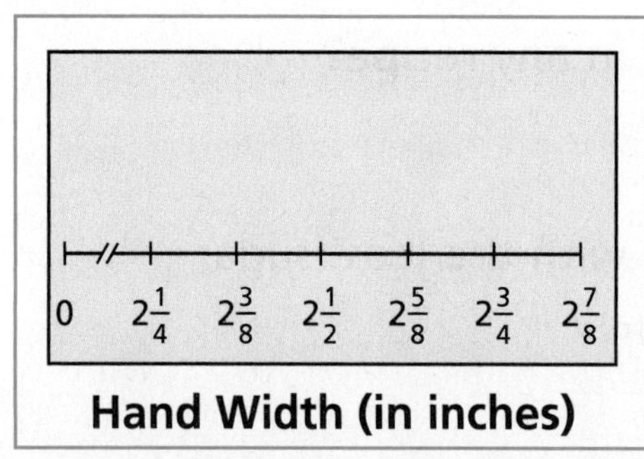

Hand Width (in inches)

10. What is the difference between the width of the widest hand and the most common hand width?

11. Write a problem you could solve by using the line plot.

Dear Family,

In this unit, your child will be introduced to decimal numbers. Students will begin by using what they already know about pennies, dimes, and dollars to see connections between fractions and decimals.

Students will explore decimal numbers by using bars divided into tenths and hundredths. They will relate decimals to fractions, which are also used to represent parts of a whole.

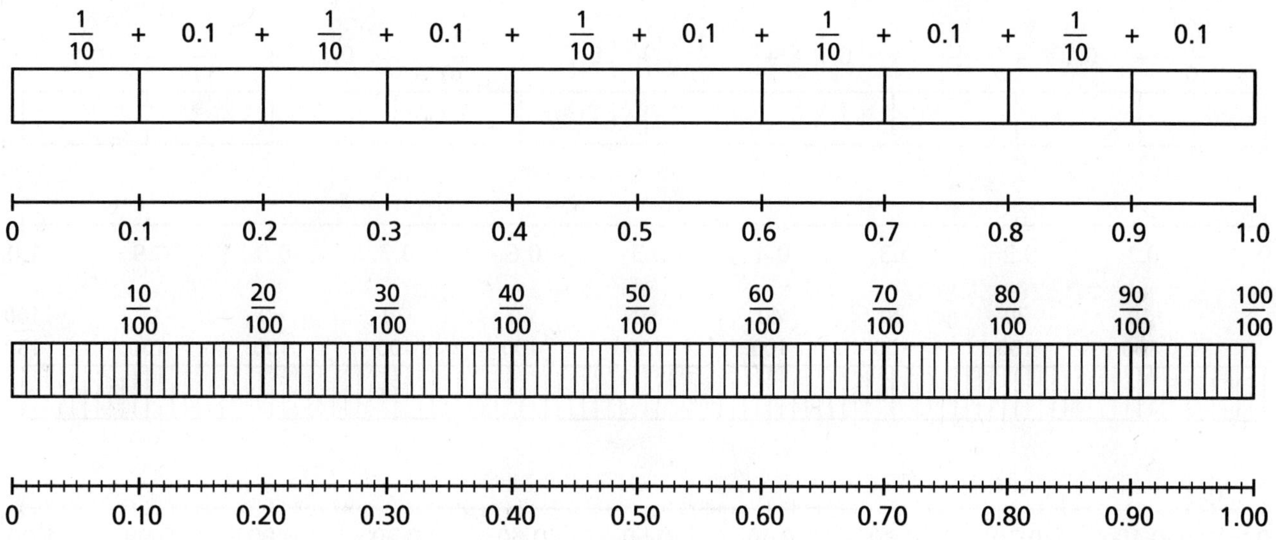

Share with your family the Family Letter on Activity Workbook page 79.

Students will read, write, and model decimal numbers. They will also learn to combine whole numbers with decimals. They will work with numbers such as 1.72 and 12.9. Students will also compare decimal numbers with other decimal numbers.

Students will apply their understanding of decimal concepts when they compare decimals.

Comparing Decimals

6.8 ◯ 3.42   6.80 (>) 3.42

Adding a zero makes the numbers easier to compare.

Please call if you have any questions or comments.

Thank you.

Sincerely,
Your child's teacher

This unit includes the Common Core Standards for Mathematical Content for Number and Operations–Fractions, and Measurement and Data, 4.NF.1, 4.NF.2, 4.NF.6, 4.NF.7, 4.MD.2, 4.MD.4, and all Mathematical Practices.

Estimada familia:

En esta unidad, se presentarán los números decimales. Para comenzar, los estudiantes usarán lo que ya saben acerca de las monedas de un centavo, de las monedas de diez y de los dólares, para ver cómo se relacionan las fracciones y los decimales.

Los estudiantes estudiarán los números decimales usando barras divididas en décimos y centésimos. Relacionarán los decimales con las fracciones que también se usan para representar partes del entero.

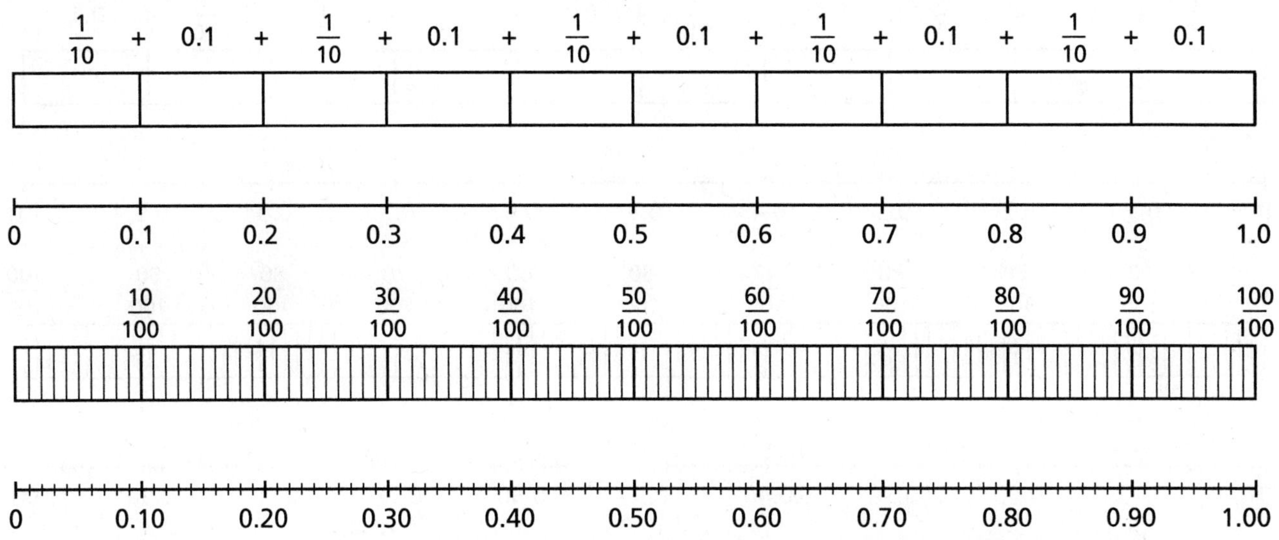

Los estudiantes leerán, escribirán y representarán números decimales. También aprenderán a combinar números enteros con decimales. Trabajarán con números tales como 1.72 y 12.9. Compararán números decimales con otros números decimales.

Al comparar decimales, los estudiantes aplicarán los conceptos decimales que ya conozcan.

Muestra a tu familia la Carta a la familia de la página 80 del Cuaderno de actividades y trabajo.

Comparar decimales

6.8 ◯ 3.42 6.80 ⓐ> 3.42

Añadir un cero facilita la comparación de números.

Si tiene alguna duda o algún comentario, por favor comuníquese conmigo.

Gracias.

Atentamente,
El maestro de su niño

Esta unidad incluye los Common Core Standards for Mathematical Content for Number and Operations–Fractions, and Measurement and Data, 4.NF.1, 4.NF.2, 4.NF.6, 4.NF.7, 4.MD.2, 4.MD.4, and all Mathematical Practices.

▶ Tenths and Hundredths

Pennies and dimes can help you understand tenths and hundredths. Discuss what you see.

100 pennies = 10 dimes = 1 dollar

100 pennies = 1 dollar   10 dimes = 1 dollar

1 penny is $\frac{1}{100}$ of a dollar   1 dime is $\frac{1}{10}$ of a dollar

1. 1 penny = $\frac{1}{100}$ = 0.01

$\frac{10}{100}$ 10 of 100 equal parts

$\frac{1}{10}$ 1 of 10 equal parts

0.1

0.10

2. 1 dime = $\frac{1}{10}$ = 0.1

$\frac{10}{100} + \frac{10}{100} = \frac{20}{100}$

$\frac{1}{10} + \frac{1}{10} = \frac{2}{10}$

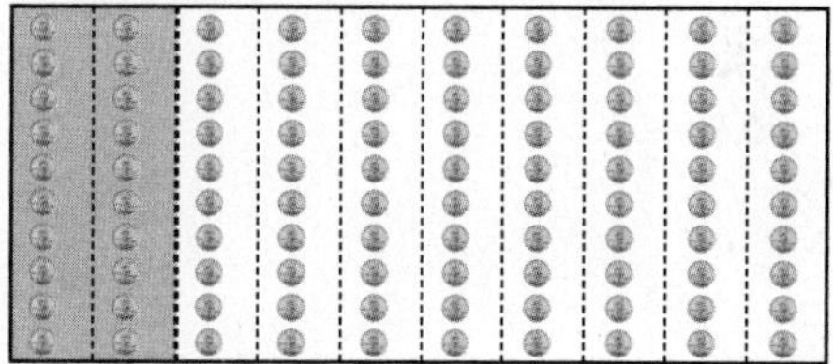

0.1 + 0.1 = 0.2

0.10 + 0.10 = 0.20

3. $\frac{10}{100} + \frac{10}{100} + \frac{5}{100} = \frac{25}{100}$

$\frac{1}{10} + \frac{1}{10} + \frac{5}{100} = \frac{25}{100}$

0.1 + 0.1 + 0.05 = 0.25

0.10 + 0.10 + 0.05 = 0.25

4. $\frac{25}{100} + \frac{25}{100} + \frac{25}{100} = \frac{75}{100}$

0.25 + 0.25 + 0.25 = 0.75

5. $\frac{1}{10} + \frac{1}{10} + \frac{1}{10} + \frac{1}{10} + \frac{1}{10} = \frac{5}{10} = \frac{1}{2}$

0.1 + 0.1 + 0.1 + 0.1 + 0.1 = 0.5

0.10 + 0.10 + 0.10 + 0.10 + 0.10 = 0.50

▶ Halves and Fourths

Equal shares of 1 whole can be written as a fraction or as a decimal. Each whole dollar below is equal to 100 pennies. Discuss the patterns you see.

6. $\frac{1}{2}$ 1 of 2 equal parts

2 equal parts

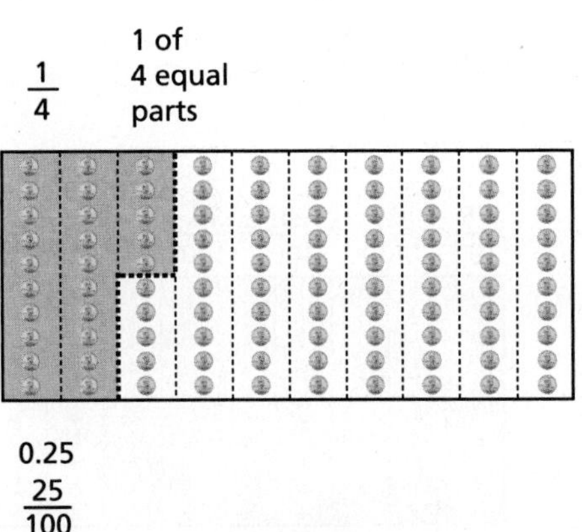

$0.5 = \frac{5}{10}$

$0.50 = \frac{50}{100}$

7. $\frac{1}{2} + \frac{1}{2} = \frac{2}{2}$ 2 of 2 equal parts = 1 whole

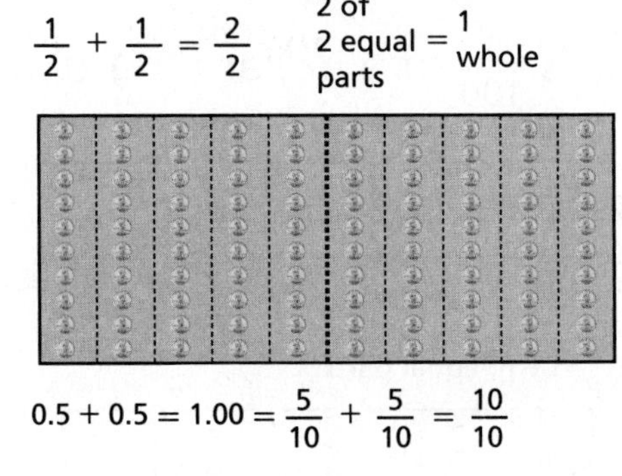

$0.5 + 0.5 = 1.00 = \frac{5}{10} + \frac{5}{10} = \frac{10}{10}$

$0.50 + 0.50 = 1.00 = \frac{50}{100} + \frac{50}{100} = \frac{100}{100}$

8. $\frac{1}{4}$ 1 of 4 equal parts

4 equal parts

0.25

$\frac{25}{100}$

9. $\frac{1}{4} + \frac{1}{4} = \frac{2}{4}$ 2 of 4 equal parts

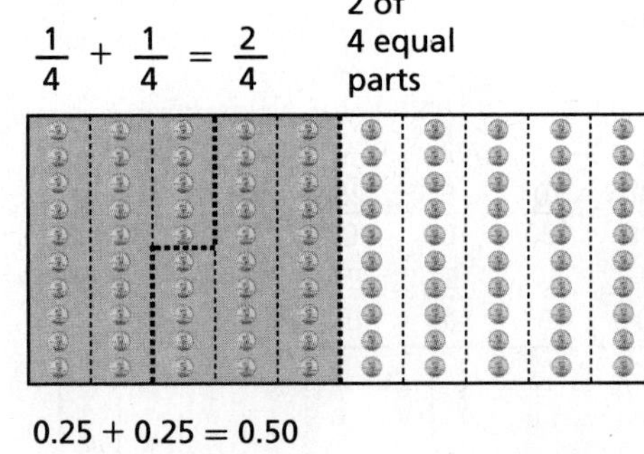

$0.25 + 0.25 = 0.50$

$\frac{25}{100} + \frac{25}{100} = \frac{50}{100}$

10. $\frac{1}{4} + \frac{1}{4} + \frac{1}{4} = \frac{3}{4}$ 3 of 4 equal parts

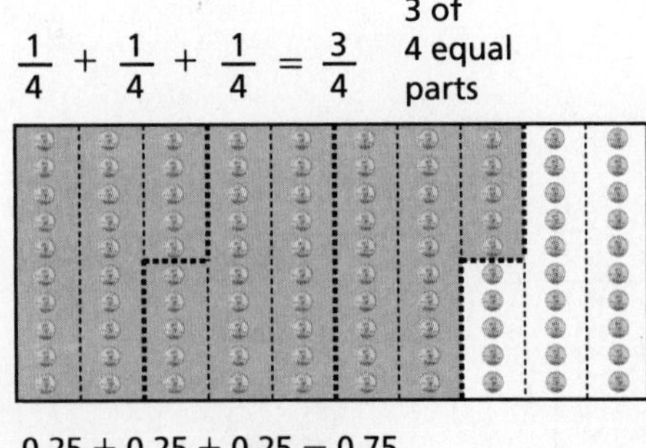

$0.25 + 0.25 + 0.25 = 0.75$

$\frac{25}{100} + \frac{25}{100} + \frac{25}{100} = \frac{75}{100}$

11. $\frac{1}{4} + \frac{1}{4} + \frac{1}{4} + \frac{1}{4} = \frac{4}{4}$ 4 of 4 equal parts = 1 whole

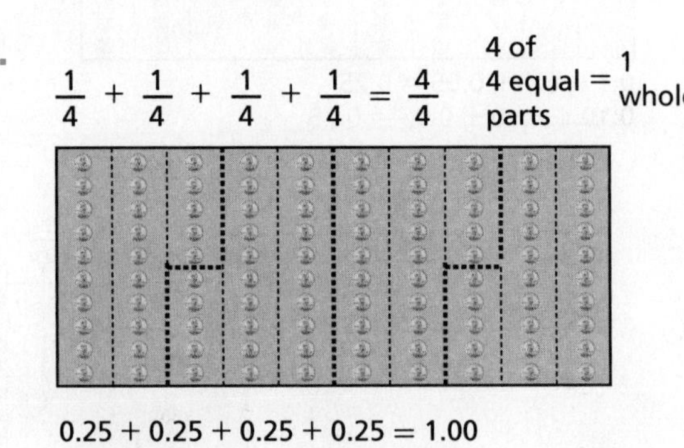

$0.25 + 0.25 + 0.25 + 0.25 = 1.00$

$\frac{25}{100} + \frac{25}{100} + \frac{25}{100} + \frac{25}{100} = \frac{100}{100}$

▶ Numbers Greater Than 1

Numbers greater than 1 can be written as fractions, decimals, or mixed numbers. A mixed number is a number that is represented by a whole number and a fraction.

Discuss the patterns you see in the equivalent fractions, decimals, and mixed numbers shown below.

12. $\frac{1}{4} + \frac{1}{4} + \frac{1}{4} + \frac{1}{4} + \frac{1}{4} = \frac{5}{4}$ 5 of 4 equal parts $= 1\frac{1}{4}$

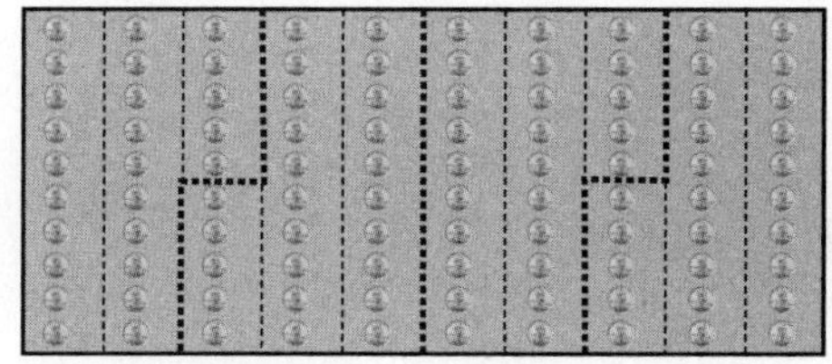

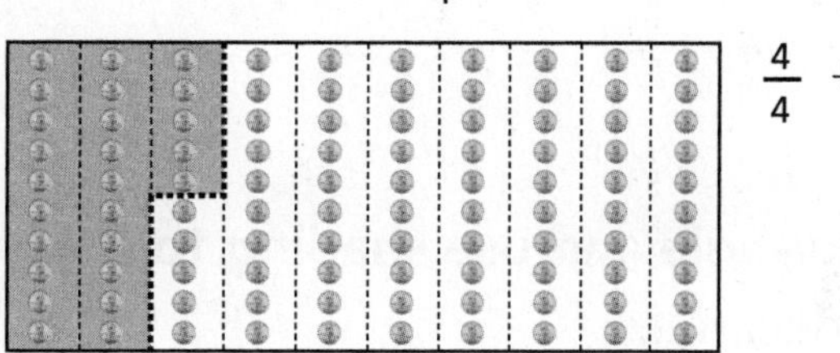

$\frac{4}{4} + \frac{1}{4}$

$0.25 + 0.25 + 0.25 + 0.25 + 0.25 = 1.25$

$\frac{25}{100} + \frac{25}{100} + \frac{25}{100} + \frac{25}{100} + \frac{25}{100} = \frac{125}{100} = 1\frac{25}{100}$

13. $\frac{1}{4} + \frac{1}{4} + \frac{1}{4} + \frac{1}{4} + \frac{1}{4} + \frac{1}{4} = \frac{6}{4}$ 6 of 4 equal parts $= 1\frac{2}{4}$

$\frac{4}{4} + \frac{2}{4}$

$0.25 + 0.25 + 0.25 + 0.25 + 0.25 + 0.25 = 1.50$

$\frac{25}{100} + \frac{25}{100} + \frac{25}{100} + \frac{25}{100} + \frac{25}{100} + \frac{25}{100} = \frac{150}{100} = \frac{100}{100} + \frac{50}{100} = 1 + \frac{50}{100} = 1\frac{50}{100}$

14. $\frac{1}{4} + \frac{1}{4} + \frac{1}{4} + \frac{1}{4} + \frac{1}{4} + \frac{1}{4} + \frac{1}{4} = \frac{7}{4}$ 7 of 4 equal parts $= 1\frac{3}{4}$

$\frac{4}{4} + \frac{3}{4}$

$0.25 + 0.25 + 0.25 + 0.25 + 0.25 + 0.25 + 0.25 = 1.75$

$\frac{25}{100} + \frac{25}{100} + \frac{25}{100} + \frac{25}{100} + \frac{25}{100} + \frac{25}{100} + \frac{25}{100} = \frac{175}{100} = \frac{100}{100} + \frac{75}{100} = 1 + \frac{75}{100} = 1\frac{75}{100}$

Use Activity Workbook page 81.

► Model Equivalent Fractions and Decimals

Write a fraction and a decimal to represent the shaded part of each whole.

15.

16.

Divide each whole and use shading to show the given fraction or decimal.

17. 0.75

18. 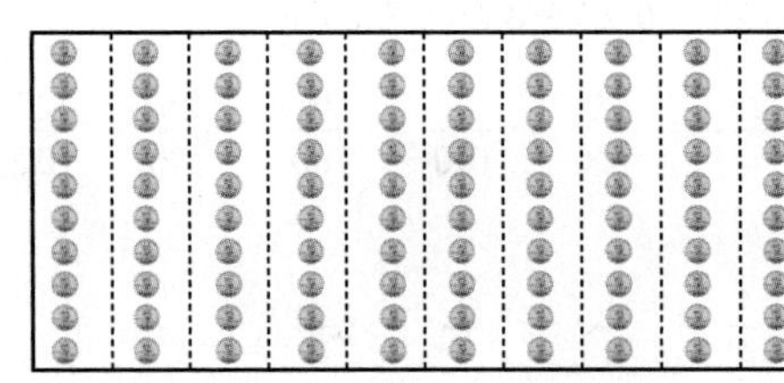 $\frac{9}{10}$

Shade these grids to show that $\frac{3}{2} = 1\frac{1}{2}$.

19. 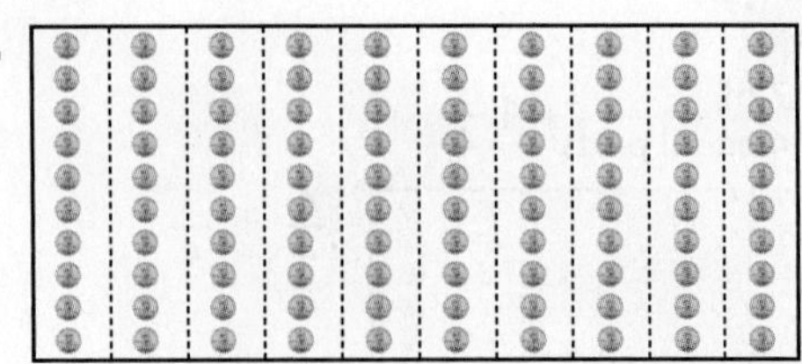

VOCABULARY
tenths
hundredths
decimal number

▶ Understand Tenths and Hundredths

Answer the questions about the bars and number lines below.

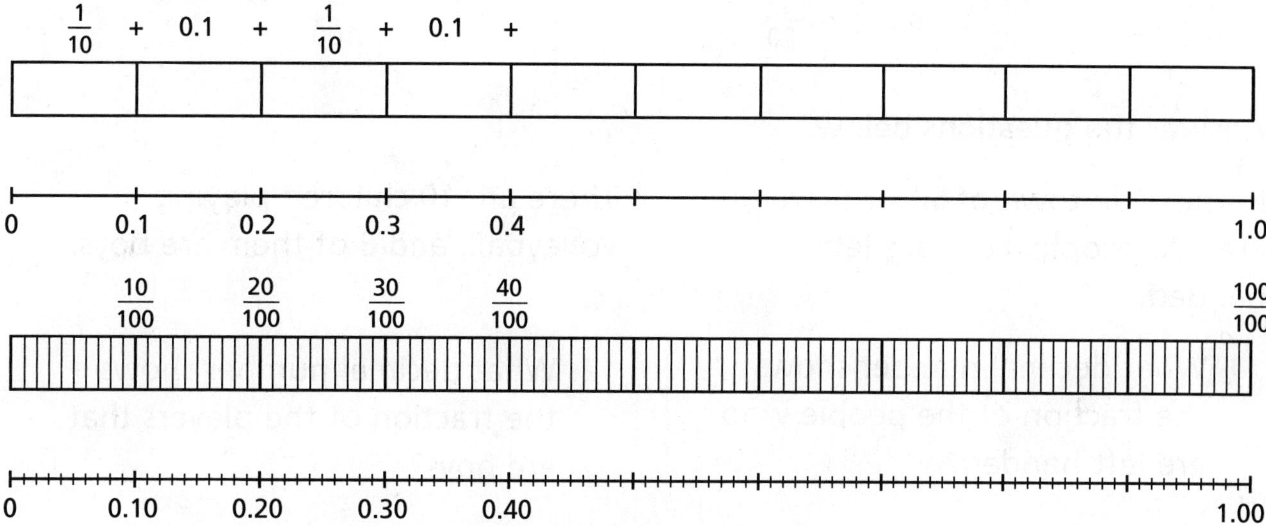

1. The bars show **tenths** and **hundredths**. Finish labeling the bars and number lines using fractions and **decimal numbers**.

2. Use what you know about fractions and about money (a dime = one tenth of a dollar and a penny = one hundredth of a dollar) to explain why 3 tenths is the same as 30 hundredths.

3. Tenths are greater than hundredths even though 10 is less than 100. Explain why this is true.

7-9
Class Activity

▶ Practice Writing Decimal Numbers

Write these numbers in decimal form.

4. 8 tenths

5. 6 hundredths

6. 35 hundredths

7. $\frac{92}{100}$

8. $\frac{2}{10}$

9. $\frac{9}{100}$

Answer the questions below.

In the little town of Silver there are 100 people. Four are left-handed.

10. What decimal number shows the fraction of the people who are left-handed?

11. What decimal number shows the fraction of the people who are right-handed?

There are 10 children playing volleyball, and 6 of them are boys.

12. What decimal number shows the fraction of the players that are boys?

13. What decimal number shows the fraction of the players that are girls?

Complete the table.

	Name of Coin	Fraction of a Dollar	Decimal Part of a Dollar
14.	Penny	$\frac{■}{100}$	
15.	Nickel	$\frac{■}{100}$ = ■	
16.	Dime	$\frac{■}{100}$ = ■	
17.	Quarter	$\frac{■}{100}$ = ■	

► Write Decimal Numbers

In the situations below, each person is traveling the same distance. Write a decimal number to represent the distance each person has traveled.

1. Aki has traveled 3 tenths of the distance, and Steven has traveled 5 tenths of the distance.

   Aki　　　　Steven

2. Jose has traveled 25 hundredths of the distance, and Lakisha has traveled 18 hundredths of the distance.

   Jose　　　　Lakisha

3. Yasir has traveled 7 tenths of the distance, and Danielle has traveled 59 hundredths of the distance.

   Yasir　　　　Danielle

4. Lea has traveled 8 hundredths of the distance, and Kwang-Sun has traveled 6 tenths of the distance.

   Lea　　　　Kwang-Sun

► Practice Comparing

Write $>$, $<$, or $=$ to compare these numbers.

5. 0.4 　 0.04	6. 0.30 　 0.3	7. 0.7 　 0.24	8. 0.1 　 0.8
9. 0.61 　 0.8	10. 0.54 　 0.2	11. 0.11 　 0.15	12. 0.02 　 0.2
13. 0.5 　 0.50	14. 0.77 　 0.3	15. 0.06 　 0.6	16. 0.9 　 0.35
17. 0.4 　 0.7	18. 0.1 　 0.10	19. 0.5 　 0.81	20. 0.41 　 0.39
21. 0.9 　 0.09	22. 0.48 　 0.6	23. 0.53 　 0.4	24. 0.70 　 0.7

▶ Word Problems With Decimal Numbers

Solve.

The Cruz family is enjoying a 10-day vacation. So far, they have been vacationing for one week.

25. What decimal number represents the part of their vacation that is past?

26. What decimal number represents the part of their vacation that remains?

Jeremy spent 3 quarters and 1 nickel at the school bookstore.

27. What decimal part of a dollar did he spend?

28. What decimal part of a dollar did he not spend?

Dana is planning to run 1 tenth of a mile every day for 8 days.

29. What part of a mile will she have run at the end of the eighth day?

30. What part of a mile will she run all together on the next two days?

▶ Practice Writing Decimal Numbers

Write the word name of each number.

31. 0.1

32. 0.73

33. 0.09

34. 0.5

Write a decimal number for each word name.

35. fourteen hundredths

36. two tenths

37. eight tenths

38. six hundredths

Use Activity Workbook page 85.

▶ Discuss Symmetry Around the Ones

× 10 (Greater) ÷ 10 (Lesser)

Hundreds	Tens	ONES	Tenths	Hundredths
100.	10.	1.	0.1	0.01
$\frac{100}{1}$	$\frac{10}{1}$	$\frac{1}{1}$	$\frac{1}{10}$	$\frac{1}{100}$
$100.00	$10.00	$1.00	$0.10	$0.01

1. Discuss symmetries and relationships you see in the place value chart.

2. Is it easier to see place value patterns in **a** or **b**? Discuss why.

   a. 500 50 5 .5 .05

   b. 500 50 5 0.5 0.05

▶ Show and Read Decimal Numbers

Use your Decimal Secret Code Cards to make numbers on the frame.

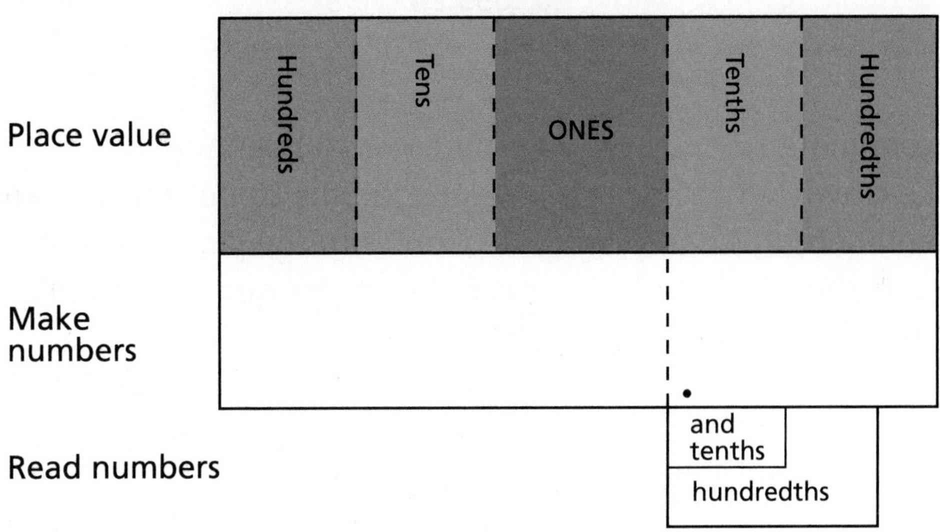

Use Activity Workbook page 86.

▶ Write Numbers in Decimal Form

Read and write each mixed number as a decimal.

3. $3\frac{1}{10}$
4. $5\frac{7}{100}$
5. $2\frac{46}{100}$
6. $28\frac{9}{10}$

Read and write each decimal as a mixed number.

7. 12.8
8. 3.05
9. 4.85
10. 49.7

Read each word name. Then write a decimal for each word name.

11. sixty-one hundredths
12. six and fourteen hundredths
13. seventy and eight tenths
14. fifty-five and six hundredths

▶ Expanded Form

Write each decimal in expanded form.

15. 8.2
16. 17.45
17. 106.24
18. 50.77
19. 312.09
20. 693.24

Solve.

21. There are 100 centimeters in 1 meter. A snake crawls 3 meters and 12 more centimeters. What decimal represents the number of meters the snake crawls?

22. There are 100 pennies in 1 dollar. A jar contains 20 dollars. You add 8 pennies to the jar. What decimal represents the number of dollars in the jar?

▶ Zeros in Greater Decimal Numbers

Use the tables to answer Problems 1–4.

Insert Zeros to the Right			
Whole Numbers		Decimal Numbers	
3	30	0.3	0.30
67	670	6.7	6.70

Insert Zeros to the Left			
Whole Numbers		Decimal Numbers	
3	03	0.3	0.03
67	067	6.7	6.07

1. What happens if we insert a zero to the right of a whole number?

2. What happens if we insert a zero to the right of a decimal number?

3. What happens if we insert a zero to the left of a whole number?

4. What happens if we insert a zero to the left of a decimal number just after the decimal point?

5. Are whole numbers and decimal numbers alike or different when it comes to putting in extra zeros? Explain your answer.

6. Do the pairs of numbers below have the same value? Why or why not?

   0.6 and .6   .25 and 0.25   0.9 and 0.90

▶ Compare Decimals

You can use your understanding of place value and the placement of zeros in decimal numbers to compare decimal numbers.

Problem:

Which of these numbers is the greatest: 2.35, 2.3, or 2.4

Solution:

2.35
2.30
(2.40)

With the places aligned and the extra zeros added, we can see which is greatest.

Write >, <, or = to compare these numbers.

7. 27.5 ● 8.37
8. 6.04 ● 5.98
9. 7.36 ● 7.38
10. 36.9 ● 37.8
11. 0.5 ● 0.26
12. 0.09 ● 0.9
13. 0.8 ● 0.80
14. 0.42 ● 0.6

Use the table to answer Problems 15 and 16.

15. Francis measured some common insects. The table shows the lengths in centimeters. List the insects from longest to shortest.

Lengths of Insects

Name	Length
Ladybug	0.64 cm
Moth	0.3 cm
Mosquito	0.32 cm
Cricket	1.8 cm
Bumblebee	2 cm

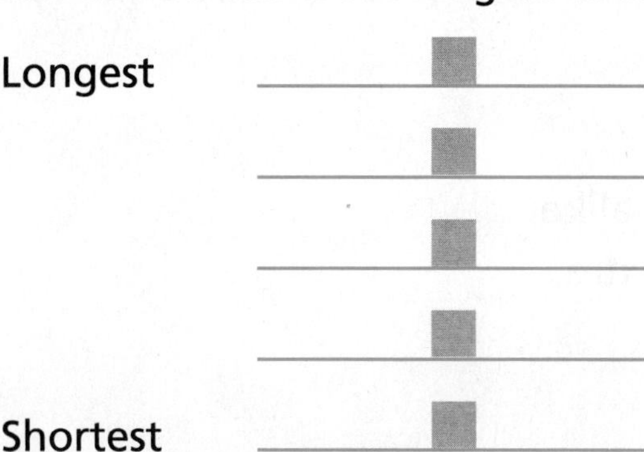

16. Maya read about a stick insect that is 1.16 centimeters long. She compared the length with the lengths in the table. Maya says the mosquito is longer than the stick insect because 0.32 > 0.16. Is Maya's reasoning correct? Explain.

▶ Math and Autumn Leaves

The weather in different parts of the United States has a noticeable effect on plants and trees. In warm parts of the country, trees can keep their leaves all year long. In the northern states, fall weather causes leaves to change color. People from around the country plan trips to see and photograph the red, yellow, orange, and brown leaves. A fall leaf-viewing trip could involve driving through a national forest, biking along a rail trail, or hiking into the mountains.

Solve.

Show your work on your paper or in your journal.

1. One popular park to photograph leaves in autumn is Macedonia Brook State Park in Kent, Connecticut. The Yellow Trail is the shortest hiking trail and is $\frac{51}{100}$ mile long. What is this fraction written as a decimal?

2. The Rogers family is visiting Massachusetts to see the leaves change color. The Old Eastern Marsh Trail is $1\frac{2}{5}$ miles long. The Bradford Rail Trail is $1\frac{3}{10}$ miles long. The Rogers family wants to take the longer trail. Which trail should they take?

Show your work on your paper or in your journal.

▶ Clarkston Park

Joshua and Lily are going north to participate in a walking tour in Clarkston Park to photograph the leaves. Here is the trail map of the different walking trails.

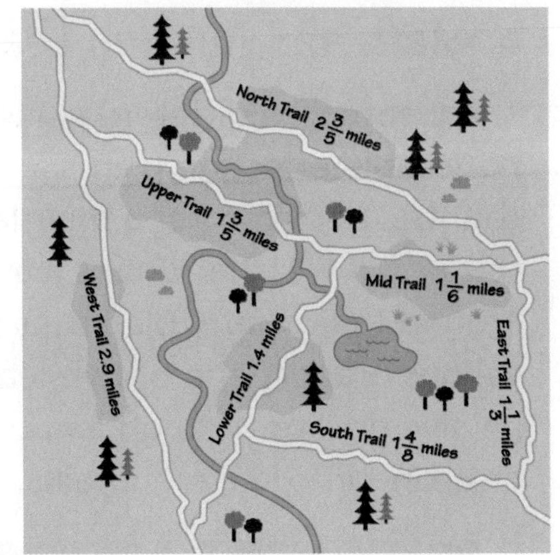

3. Write the length of the West Trail as a fraction.

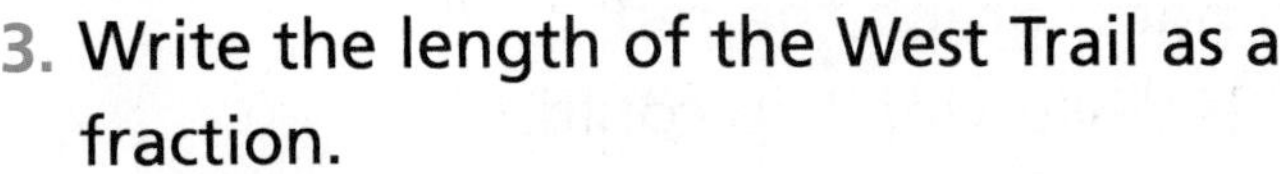

4. Which trail is shorter: the West Trail or the North Trail? Write the comparison using >, <, or =.

5. Write the length of the Lower Trail as a fraction.

6. Which trail is longer: the Lower Trail or the South Trail? Write the comparison using >, <, or =.

7. Write a fraction that is equivalent to the length of the North Trail.

8. Use the number line below and the benchmark fractions to name which trail is represented by each point.

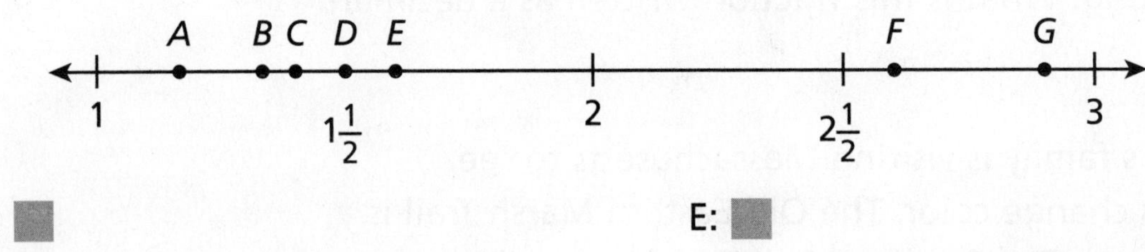

A: ▇

B: ▇

C: ▇

D: ▇

E: ▇

F: ▇

G: ▇

Use the Activity Workbook Unit Test on pages 89–92.

▶ Vocabulary

VOCABULARY
common denominator
decimal number
equivalent fractions
line plot

Choose the best term from the box.

1. ____________ are two or more fractions that represent the same part of a whole. (Lesson 7-4)

2. A graph that shows data on a number line is a ____________. (Lesson 7-7)

3. A ____________ is separated by a decimal point and shows the whole-number part on the left and the fraction part on the right. (Lesson 7-9)

▶ Concepts and Skills

4. Leonard bought a large water bottle and Natasha bought a small water bottle. They each drank $\frac{1}{2}$ of their bottles. Did they drink the same amount? Explain. (Lesson 7-3)

5. Explain how to compare $\frac{3}{4}$ and $\frac{3}{5}$. (Lesson 7-1)

6. Is $\frac{1}{4}$ equivalent to $\frac{3}{8}$? Explain. (Lesson 7-4)

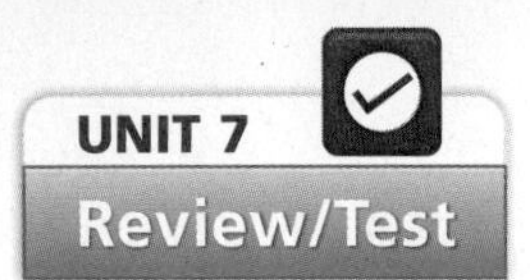

Simplify each fraction. (Lesson 7-5)

7. $\frac{9}{12} = \blacksquare$

8. $\frac{15}{40} = \blacksquare$

9. Write 5 fractions that are equivalent to $\frac{1}{5}$. (Lesson 7-4)

10. Label the point for each fraction or mixed number with the corresponding letter. (Lesson 7-2)

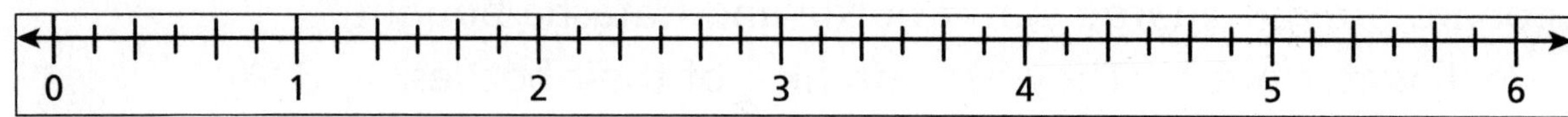

a. $1\frac{2}{3}$ b. $5\frac{1}{6}$ c. $\frac{1}{2}$ d. $3\frac{5}{6}$ e. $4\frac{1}{3}$

Which fraction is closest to 4?

Write each number in decimal form. (Lessons 7-9, 7-11)

11. seventy-four hundredths

12. $8\frac{3}{10}$

13. $12\frac{4}{100}$

14. twenty-one and thirty-five hundredths

Write >, <, or = to make each statement true. (Lessons 7-6, 7-10, 7-12)

15. $\frac{3}{4}$ ● $\frac{8}{12}$

16. $\frac{2}{3}$ ● $\frac{6}{9}$

17. 0.7 ● 0.75

18. 0.4 ● 0.25

19. 8.04 ● 8.40

20. 50.07 ● 5.70

▶ Problem Solving

Solve.

21. A farm stand sells cartons of blueberries by weight. The stand weighs each carton to determine its price. The data show the weights of the cartons of blueberries that the stand is selling. (Lesson 7-7)

a. Make a line plot to display the data.

Weight (in pounds)	Number of Cartons
$\frac{1}{4}$	2
$\frac{1}{2}$	3
$\frac{3}{4}$	5
1	3
$1\frac{1}{4}$	1

b. What general statement can you make about the weights of the blueberry cartons?

22. The shaded part of the model represents the number of pennies Nate has in his jar. Write the number of pennies Nate has in his jar as a fraction and as a decimal. (Lesson 7-8)

23. A pad of paper has 100 sheets. Helena has 4 full pads and 53 loose sheets of paper. What decimal number represents the number of sheets of paper Helena has? (Lesson 7-11)

24. Vaughn bought 10 tomato seedlings. He has 7 seedlings that are cherry tomatoes and the rest are plum tomatoes. What decimal number shows the fraction of tomatoes that are plum tomatoes? (Lesson 7-9)

25. **Extended Response** Four track team members run in the 200-meter relay race. Kaya ran 200-meters in 31.09 seconds. Both Sara and Min ran 200-meters in 31.9 seconds. Lana ran the race in 31.90 seconds. Did all of the team members run the race in the same amount of time? Explain. (Lesson 7-12)

Dear Family,

In the first half of Unit 8, your child will be learning to recognize and describe geometric figures. One type of figure is an angle. Your child will use a protractor to find the measures of angles.

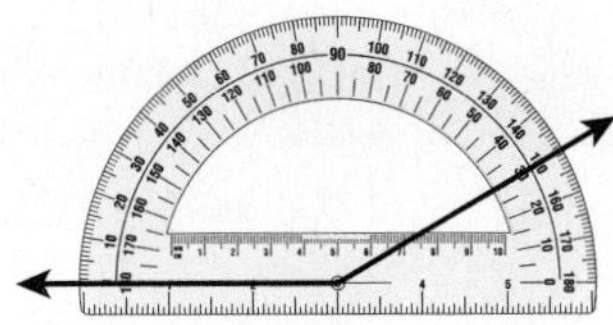

Other figures, such as triangles, may be named based on their angles and sides.

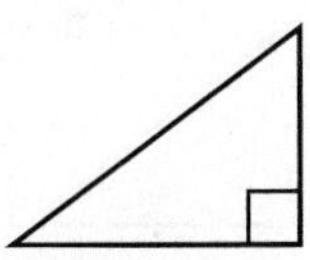

Right triangle

One right angle (90°)

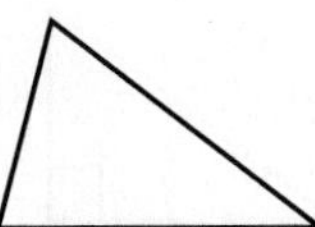

Acute triangle

All angles less than 90°

Obtuse triangle

One angle greater than 90°

Share with your family the Family Letter on Activity Workbook page 93.

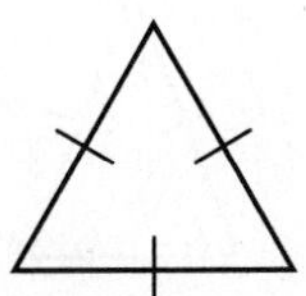

Equilateral triangle

all three sides of equal length

Isosceles triangle

two sides of equal length

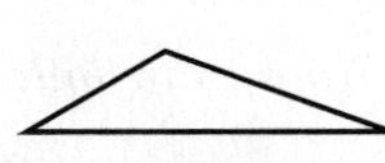

Scalene triangle

three sides of different lengths

Be sure that your child continues to review and practice the basics of multiplication and division. A good understanding of the basics will be very important in later math courses when students learn more difficult concepts in multiplication and division.

If you have any questions or comments, please call or write to me.

Thank you.

Sincerely,
Your child's teacher

This unit includes the Common Core Standards for Mathematical Content for Measurement and Data, 4.MD.5, 4.MD.5a, 4.MD.5b, 4.MD.6, 4.MD.7; Geometry, 4.G.1, 4.G.2; and all the Mathematical Practices.

Estimada familia:

En la primera parte de la Unidad 8, su niño aprenderá a reconocer y a describir figuras geométricas. Un ángulo es un tipo de figura. Su niño usará un transportador para hallar las medidas de los ángulos.

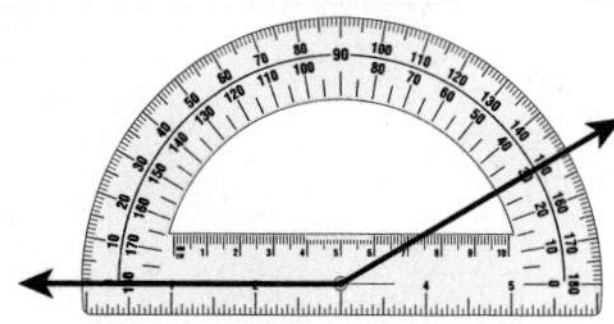

Otras figuras, tales como los triángulos, se nombran según sus ángulos y lados.

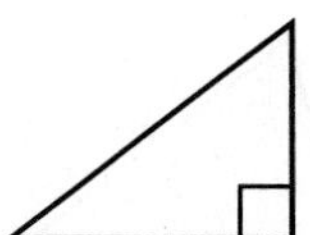

Triángulo rectángulo

Tiene un ángulo recto (90°)

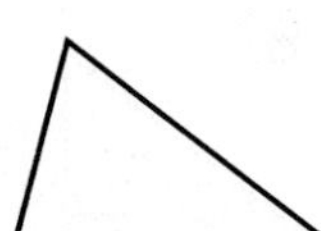

Triángulo acutángulo

Todos los ángulos son menores que 90°

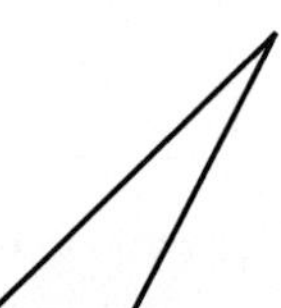

Triángulo obtusángulo

Tiene un ángulo mayor que 90°

Muestra a tu familia la Carta a la familia de la página 94 del Cuaderno de actividades y trabajo.

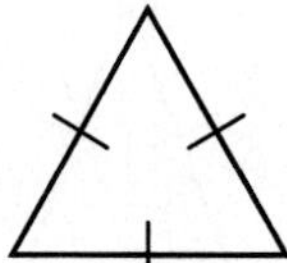

Triángulo equilátero

los tres lados tienen la misma longitud

Triángulo isósceles

dos lados tienen la misma longitud

Triángulo escaleno

los tres lados tienen diferente longitud

Asegúrese de que su niño siga repasando y practicando las multiplicaciones y divisiones básicas. Es importante que domine las operaciones básicas para que, en los cursos de matemáticas de más adelante, pueda aprender conceptos de multiplicación y división más difíciles.

Si tiene alguna pregunta o algún comentario, por favor comuníquese conmigo.

Gracias.

Atentamente,
El maestro de su niño

Esta unidad incluye los Common Core Standards for Mathematical Content for Measurement and Data, 4.MD.5, 4.MD.5a, 4.MD.5b, 4.MD.6, 4.MD.7; Geometry, 4.G.1, 4.G.2; and all the Mathematical Practices.

VOCABULARY
point
line
line segment
endpoint

▶ Points, Lines, and Line Segments

A **point** is shown by a dot. It is named by a capital letter

•*X*

A **line** is a straight path that goes on forever in both directions. When you draw a line, you put arrows on the ends to show that it goes on and on. Lines can be named by any two points on the line. Here are $\overleftrightarrow{AB}$, $\overleftrightarrow{GK}$, and $\overleftrightarrow{PN}$.

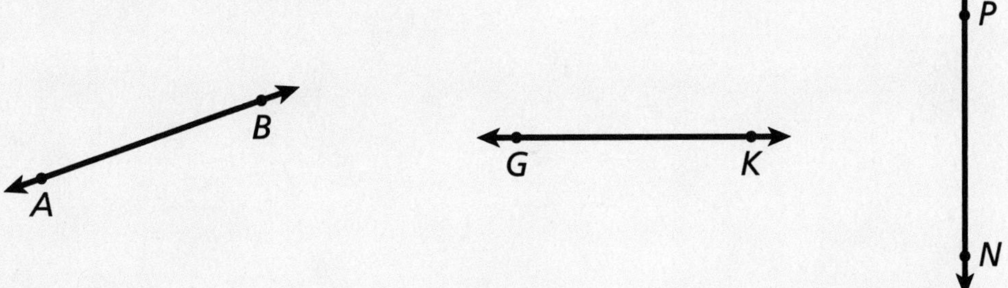

A **line segment** is part of a line. It has two ends, which are called **endpoints**. Segments are named by their endpoints. Here are $\overline{RS}$, $\overline{WT}$, and $\overline{DJ}$.

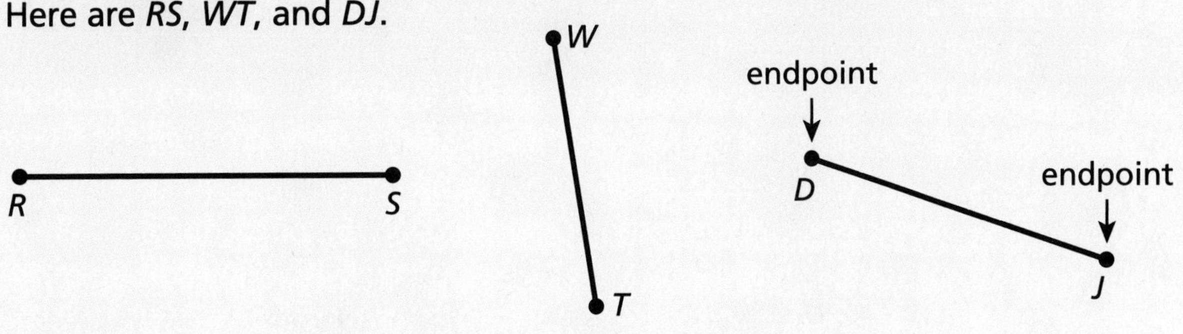

1. You can measure to find the length of a line segment, but you cannot measure to find the length of a line. Explain why.

Use Activity Workbook page 95.

VOCABULARY
angle
ray
vertex

► Drawing Points, Rays, and Angles

An **angle** is formed by two **rays** with the same endpoint, called the **vertex**.

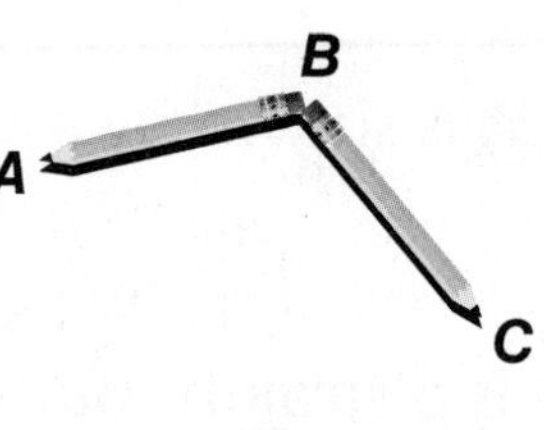

You can label figures with letters to name them. This is $\angle ABC$. Its rays are $\overrightarrow{BA}$ and $\overrightarrow{BC}$.

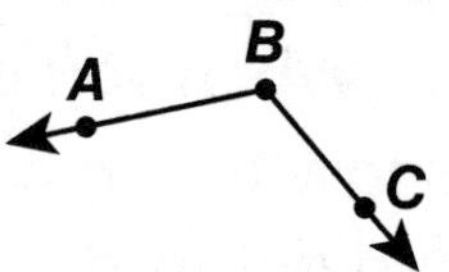

Draw and label each figure.

2. Draw and label a point. Write the name of your point.

3. Draw a ray. Label the endpoint. Write the name of your ray.

4. Draw an angle. Label the vertex and the two rays. Write the name of your angle.

VOCABULARY
acute angle
right angle
obtuse angle

▶ Discuss Angles

Angles can be many different sizes.

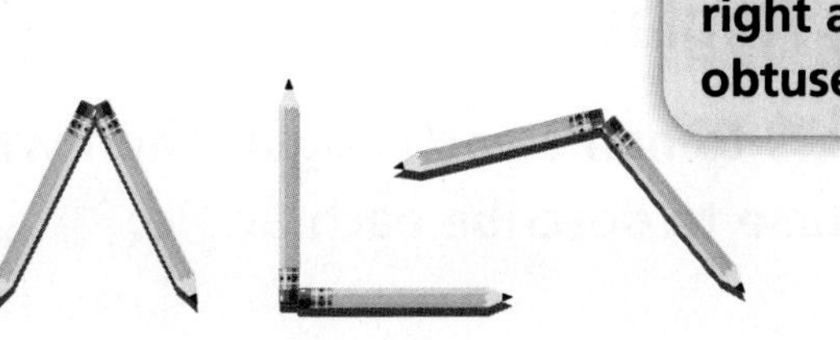

Discuss the groups of angles.

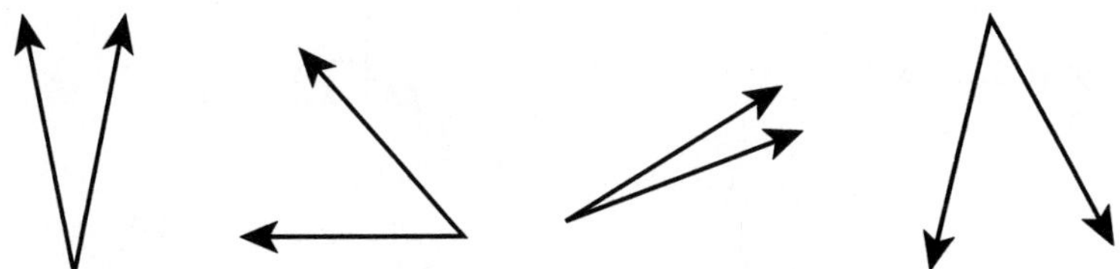

5. How are all of these **acute angles** alike?

6. How is an acute angle different from a **right angle**?

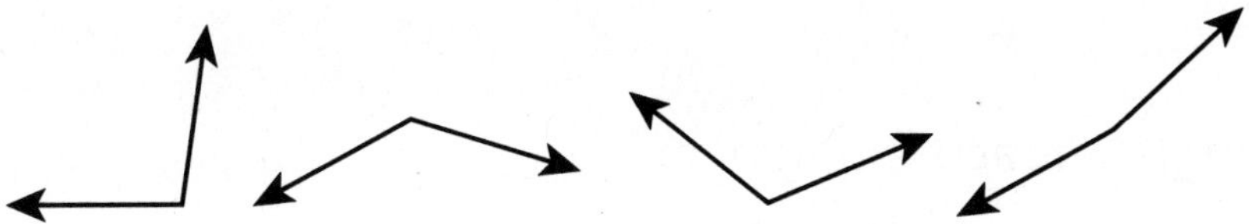

7. How are all of these **obtuse angles** alike?

8. How is an obtuse angle different from a right angle?

9. How is an obtuse angle different from an acute angle?

Use Activity Workbook page 96.

► Classify Angles

Use the letters to name each angle. Then write *acute*, *right*, or *obtuse* to describe each angle.

10. 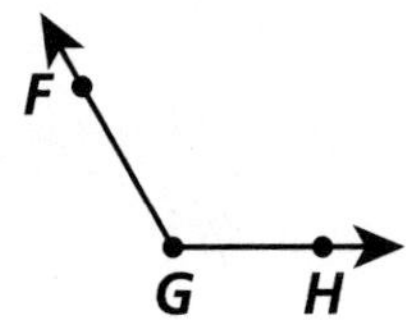

11. 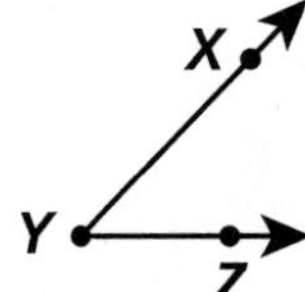

12. 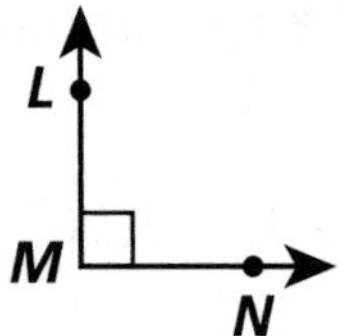

13. Use the letters to name two acute and two obtuse angles in this figure. Write *acute* or *obtuse* to describe each angle.

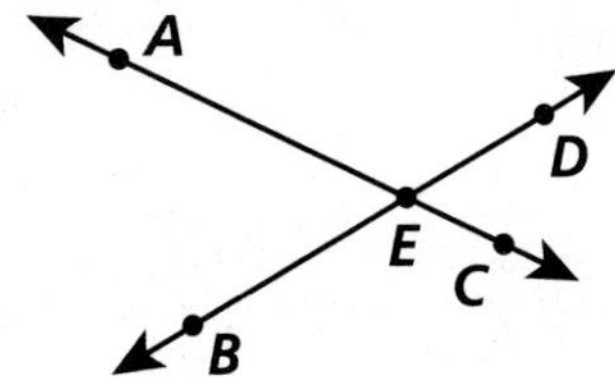

14. Draw and label a right angle, an acute angle, and an obtuse angle.

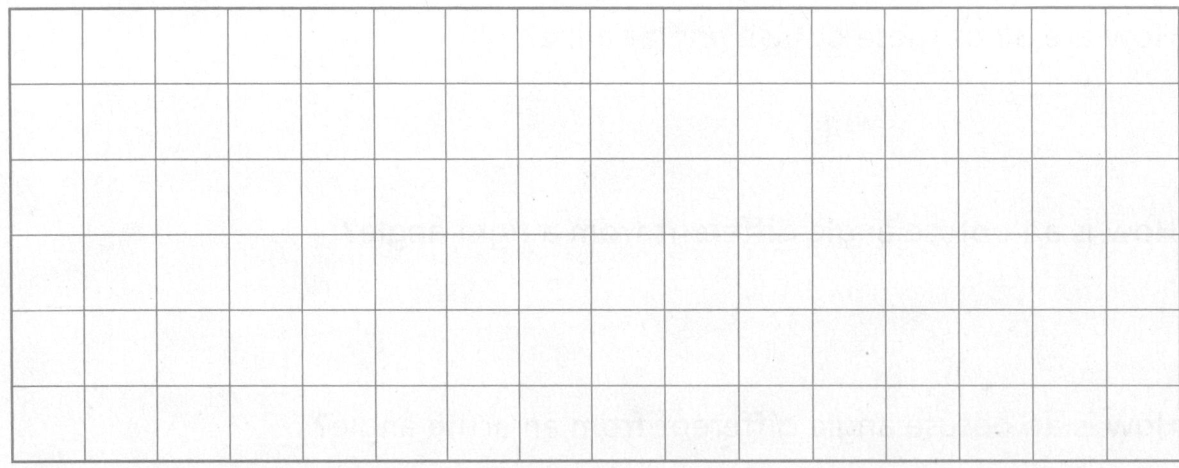

VOCABULARY
degree
right angle
straight angle

▶ Introduce Degrees

Angles are measured in units called **degrees**. One degree is the measure of one very small turn from one ray to the other.

This angle has a measure of 1 degree.

The measure of an angle is the total number of 1-degree angles that fit inside it.

This angle measures 5 degrees.

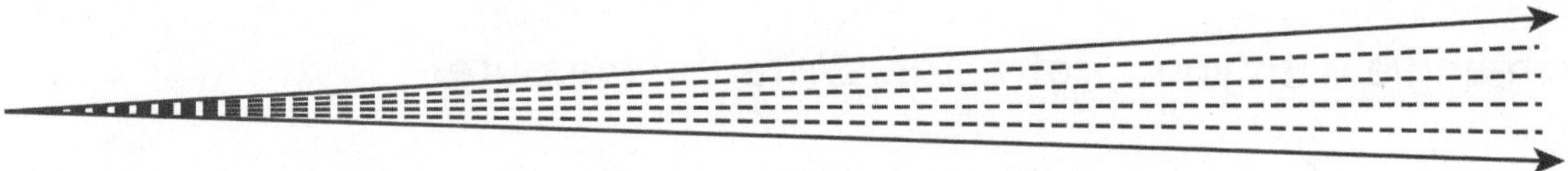

The symbol for degrees is a small raised circle (°). You can write the measure of the angle above as 5°. $5 \times 1° = 5°$

A **right angle** has a measure of 90°. A 90° turn traces one quarter of a circle.

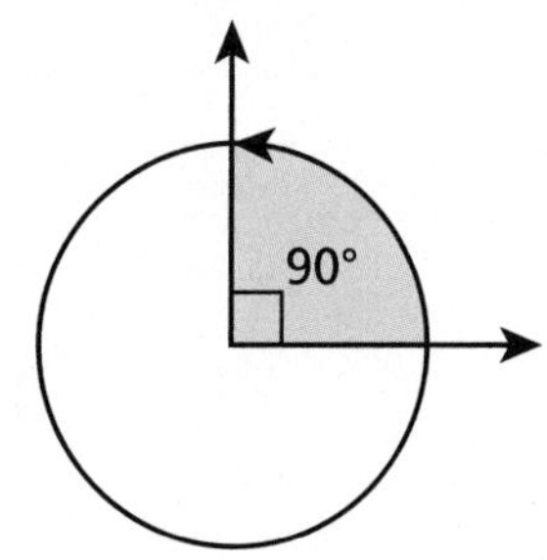

A **straight angle** measures 180°. A 180° turn traces one half of a circle.

The angle below measures 360°. A 360° turn traces a complete circle.

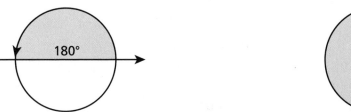

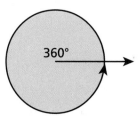

VOCABULARY
protractor

► Use a Protractor

A **protractor** is a tool that is used to measure angles in degrees. This protractor shows that ∠*ABC* measures 90°.

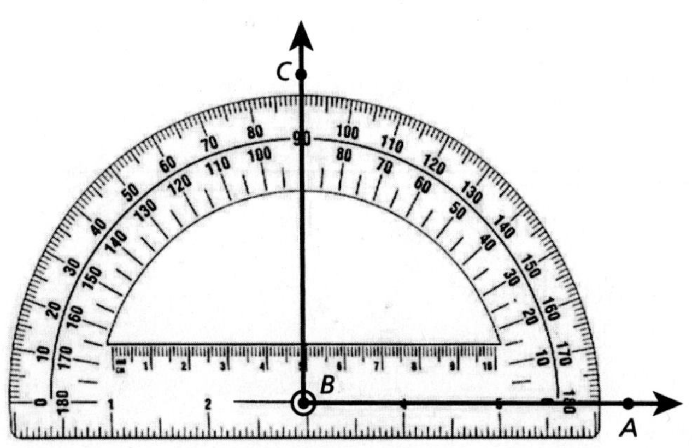

Measure each angle with your protractor. Write the measure.

1.

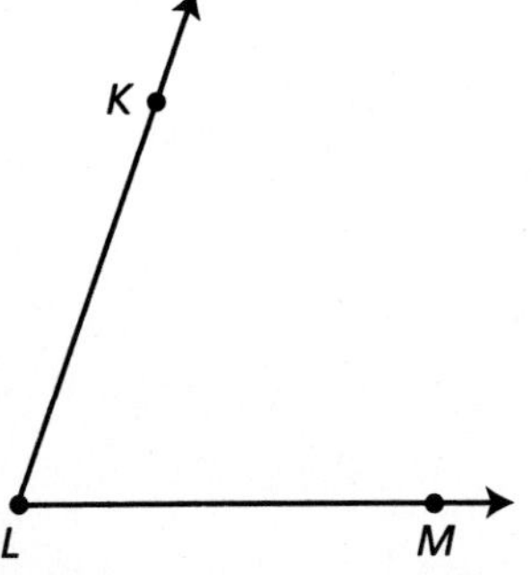

∠*KLM* = ▇

2.

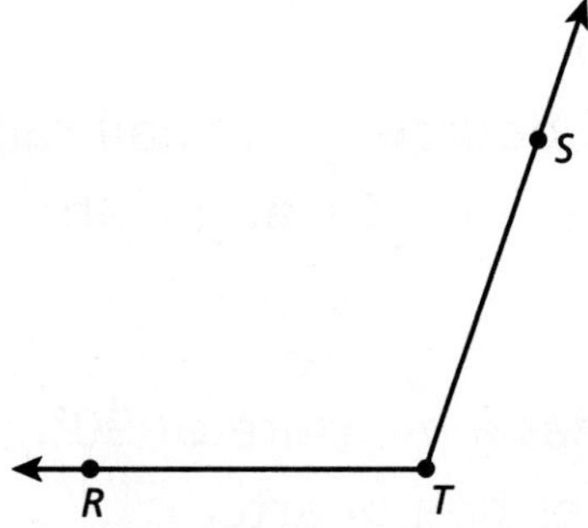

∠*STR* = ▇

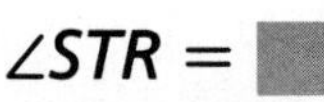

3.

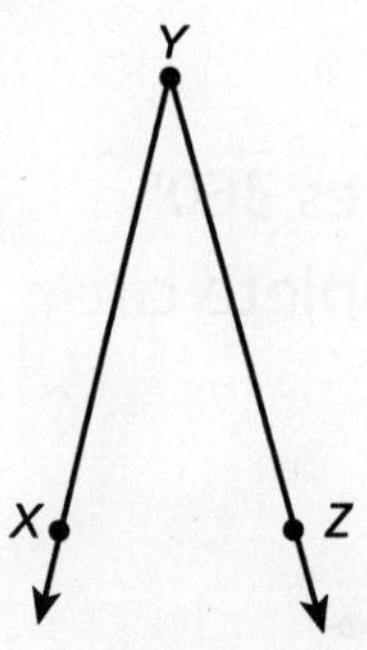

∠*XYZ* = ▇

4.

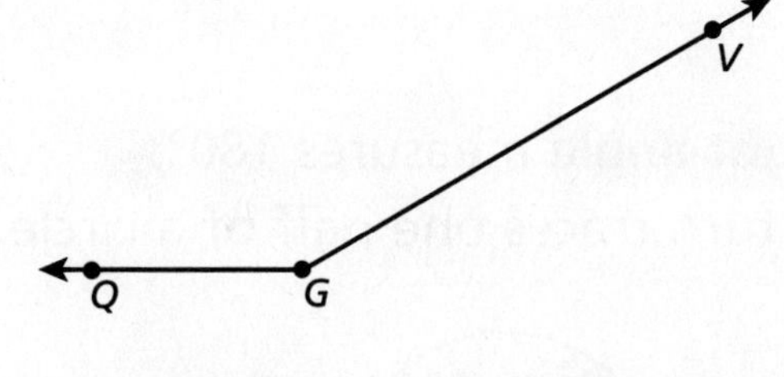

∠*QGV* = ▇

▶ Sketch Angles

Sketch each angle, or draw it using a protractor.

5. 90°

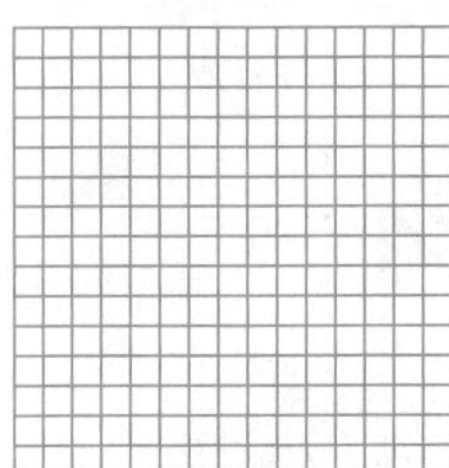

6. 45°

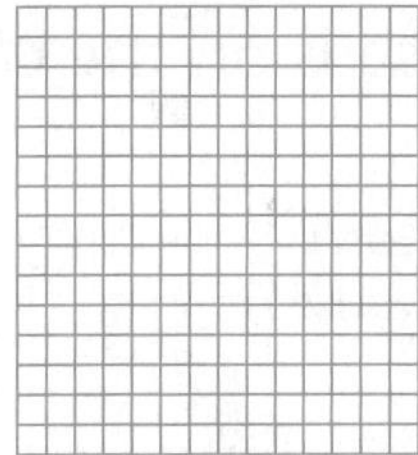

7. 180°

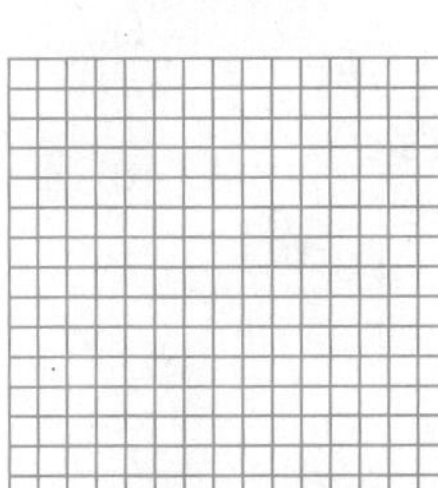

8. 360°

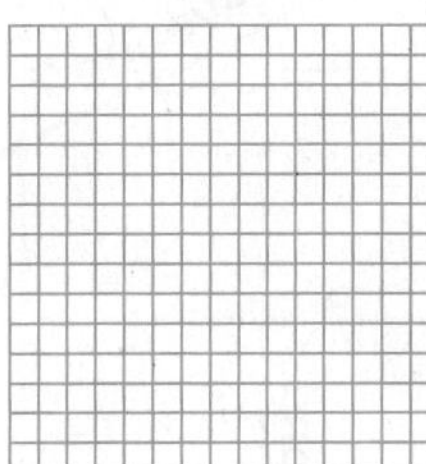

▶ Use Reasoning

Use the figures at the right to answer the following questions.

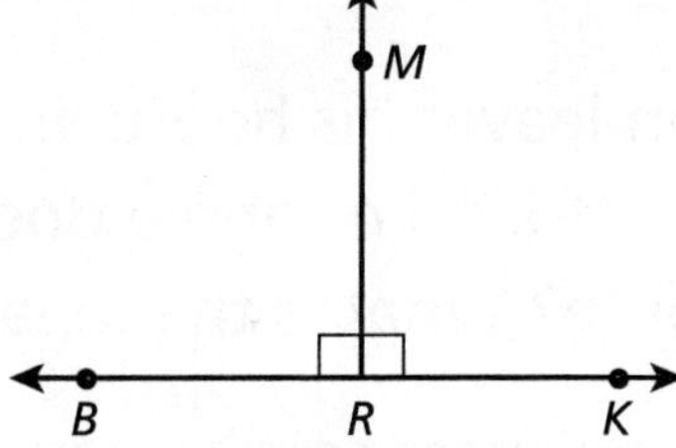

9. Name one right angle in each figure.

10. Name one straight angle in each figure.

11. How much greater is the measure of $\angle KRB$ than the measure of $\angle IAO$?

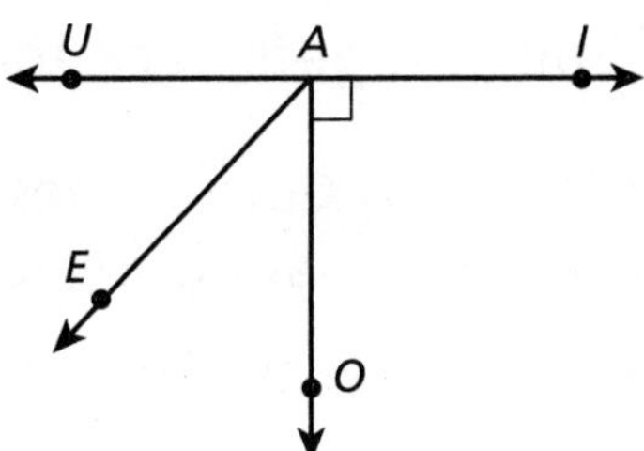

12. Which angle appears to be a 45° angle?

13. The measure of $\angle IAE$ is 135°.

What is the measure of $\angle OAE$?

What is the measure of $\angle UAE$?

► Angles in the Real World

Here is a map of Jon's neighborhood. The east and west streets are named for presidents of the United States. The north and south streets are numbered. The avenues have letters. Jon's house is on the corner of Lincoln and First.

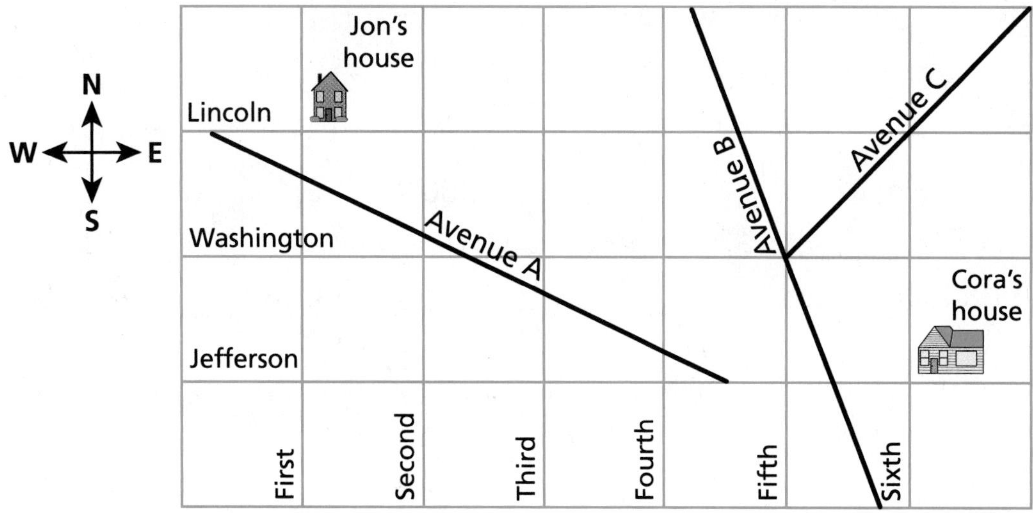

14. What do the arrows to the left of the map tell you?

15. Jon leaves his house and rides his bike south on First. What kind of angle does he make for each turn in this route? What is the measure of each angle?

- Jon turns southeast onto Avenue A.
- When he reaches Washington, he turns west.
- When he gets back to First, he turns south.

16. Jon's cousin Cora leaves Jon's house and rides east on Lincoln to Avenue B. Draw the angle Cora makes if she turns southeast. What is the measure of the angle?

VOCABULARY
circle
reflex angle

► Measure Angles in a Circle

You can show all the different types of angles in a circle.

Acute angle

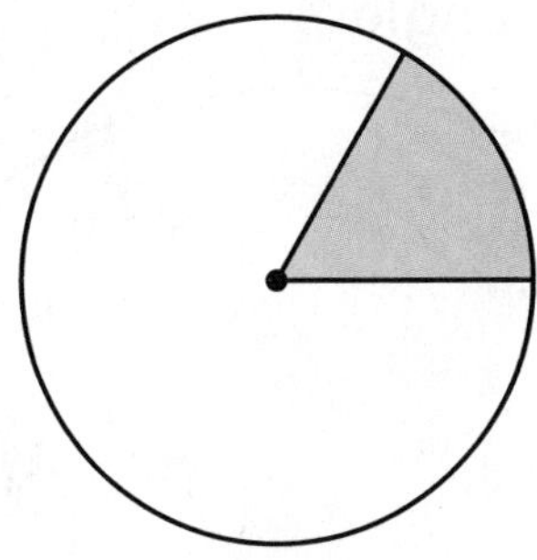

greater than 0° and less than 90°

Right angle

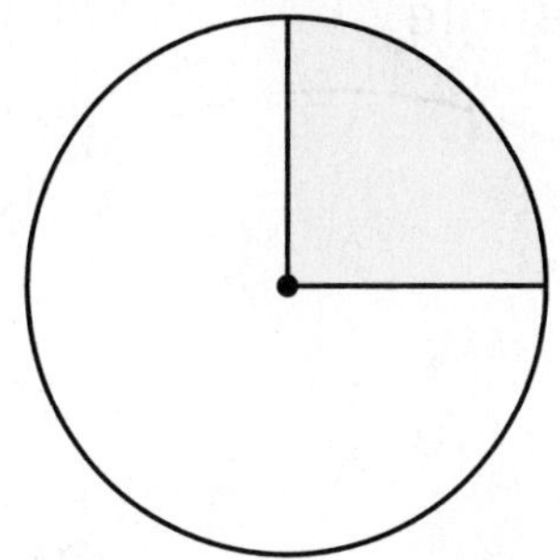

90°

Obtuse angle

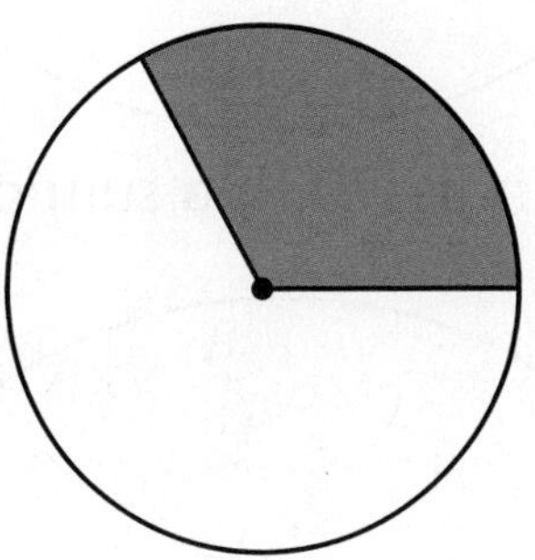

greater than 90° and less than 180°

Straight angle

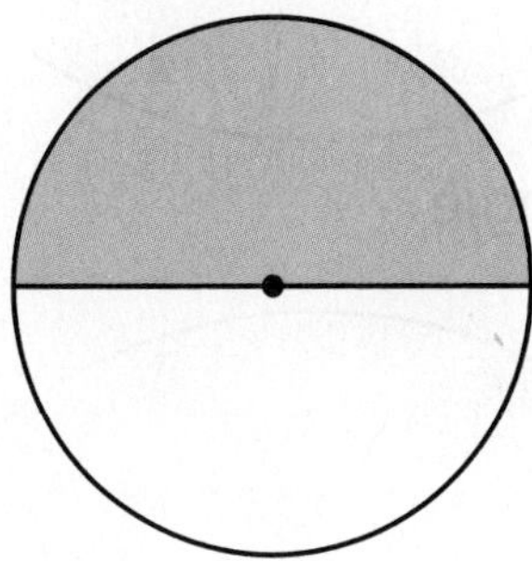

180°

Reflex angle

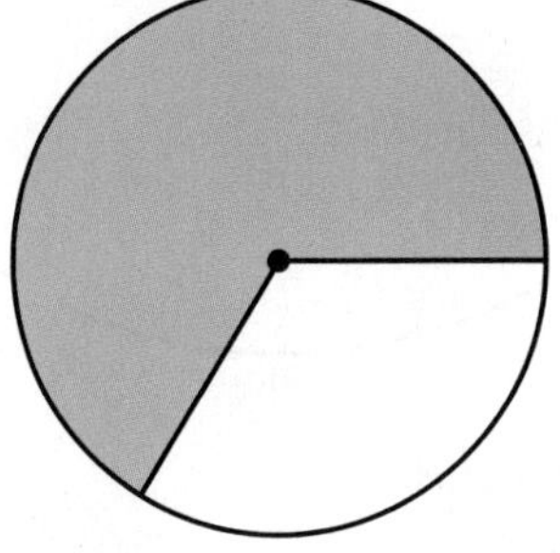

greater than 180° and less than 360°

Circle

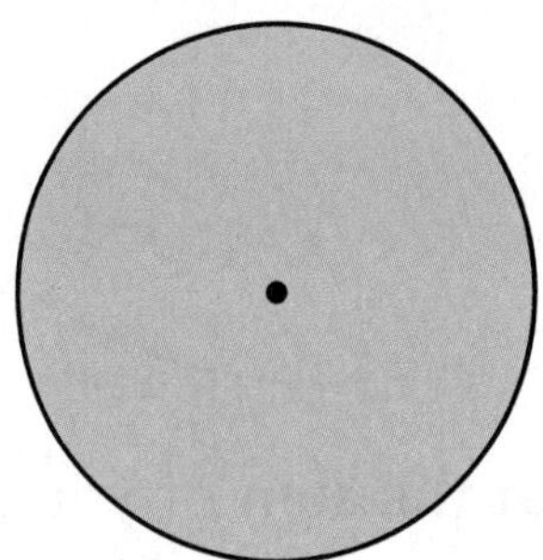

360°

► Draw Angles in a Circle

Use a straightedge and a protractor to draw and shade an angle of each type. Measure and label each angle.

1. obtuse angle

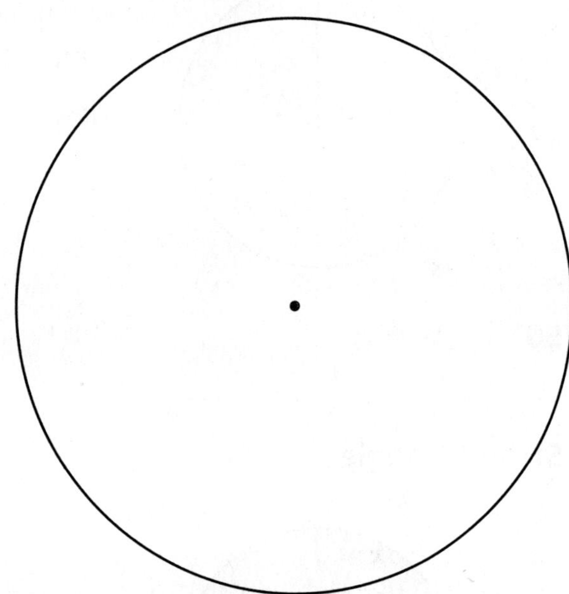

2. straight angle

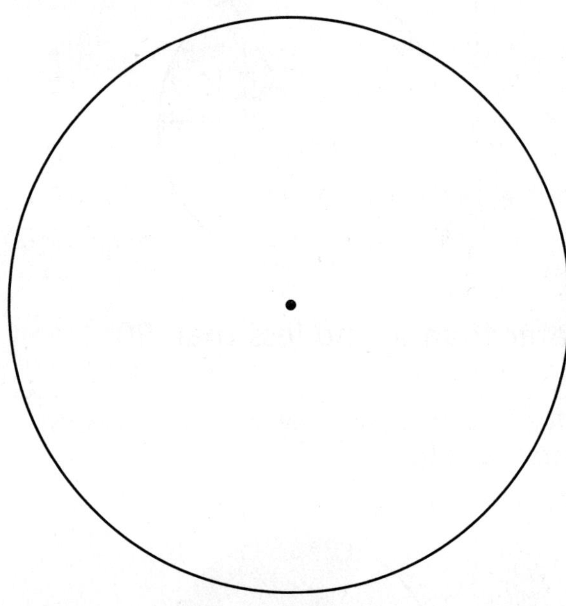

3. acute angle

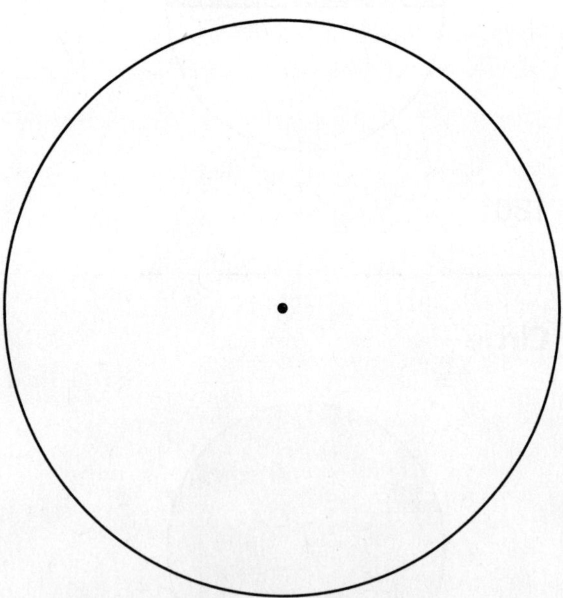

4. three angles with a sum of 360°

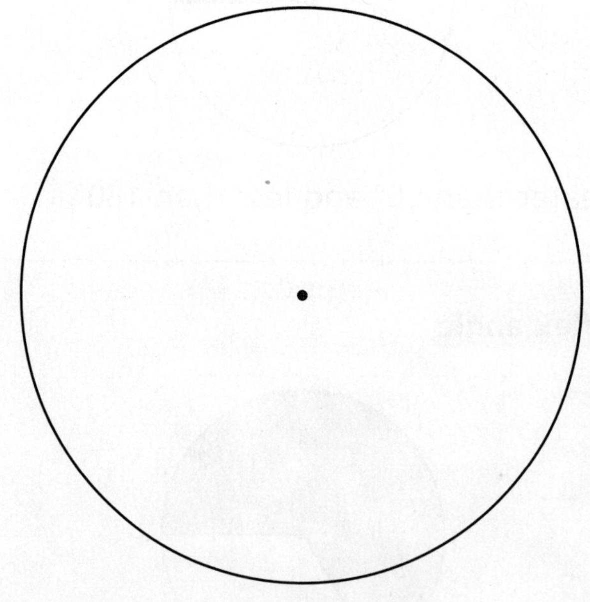

5. Write out the sum of your angle measures in Exercise 4 to show that it equals 360°

Use Activity Workbook page 102.

VOCABULARY
right triangle
obtuse triangle
acute triangle

► Discuss Angles of a Triangle

The prefix *tri-* means "three," so it is easy to remember that a triangle has 3 angles. Triangles can take their names from the kind of angles they have.

- A **right triangle** has one right angle, which we show by drawing a small square at the right angle.
- An **obtuse triangle** has one obtuse angle.
- An **acute triangle** has three acute angles.

1. You can also use letters to write and talk about triangles. This triangle is $\triangle QRS$. Name its three angles and their type.

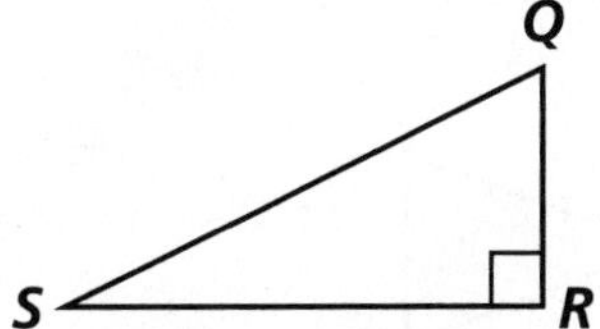

2. What kind of triangle is $\triangle QRS$? How do you know?

3. Draw and label a right triangle, an acute triangle, and an obtuse triangle.

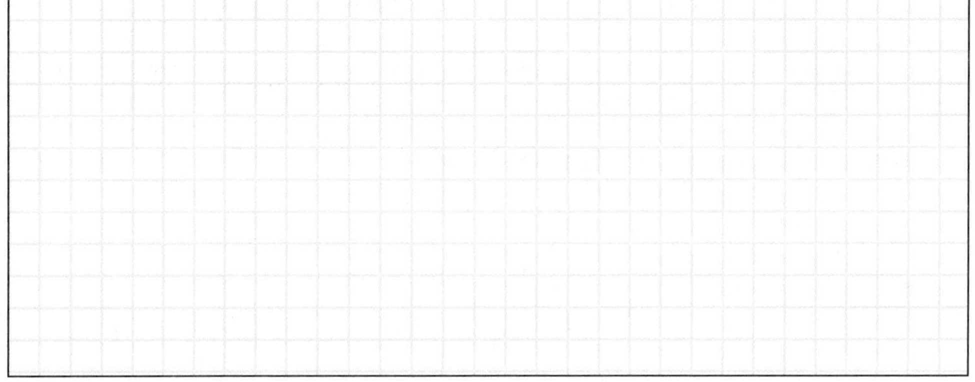

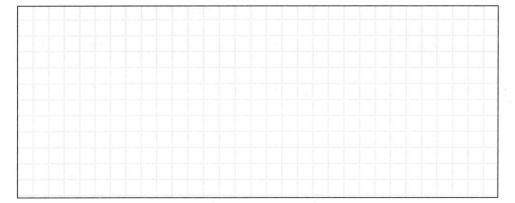

Show your work on your paper or in your journal.

► Identify Angles of a Triangle

Name each triangle by its angles. Explain your thinking.

4.

5.

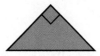

6.

7. 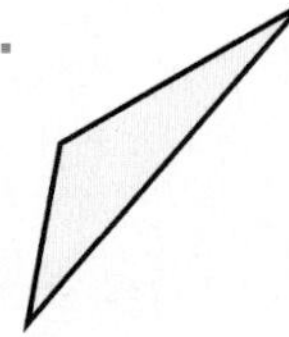

8.

9. 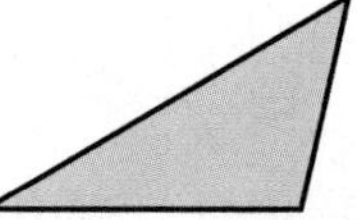

10. 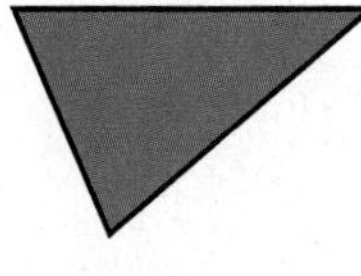

11. 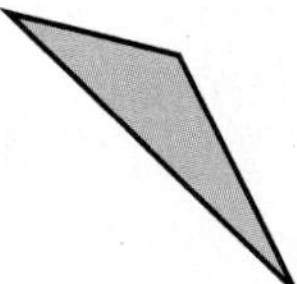

12. 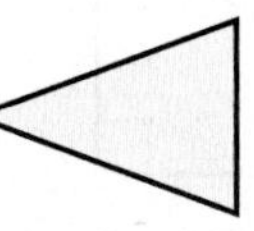

13. 

14. 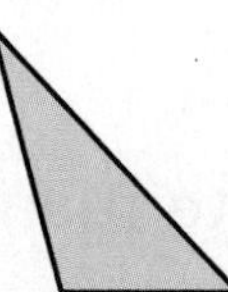

15. 

16. Describe how angles make triangles different from one another.

VOCABULARY
equilateral
isosceles
scalene

▶ Discuss Sides of a Triangle

Triangles can be named by their sides. Small "tick marks" on the sides of triangles tell us when sides are equal.

- The prefix *equi-* means "equal." Triangles that have three equal sides are called **equilateral**.

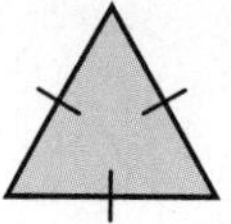

- Triangles that have two equal sides are called **isosceles**. The word *isosceles* comes from very old words that mean "equal legs."

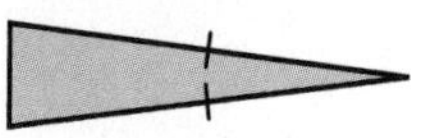

- Triangles with no equal sides are called **scalene**. All triangles that are not equilateral or isosceles are scalene.

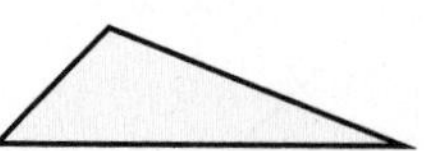

Use these triangles to answer the questions.

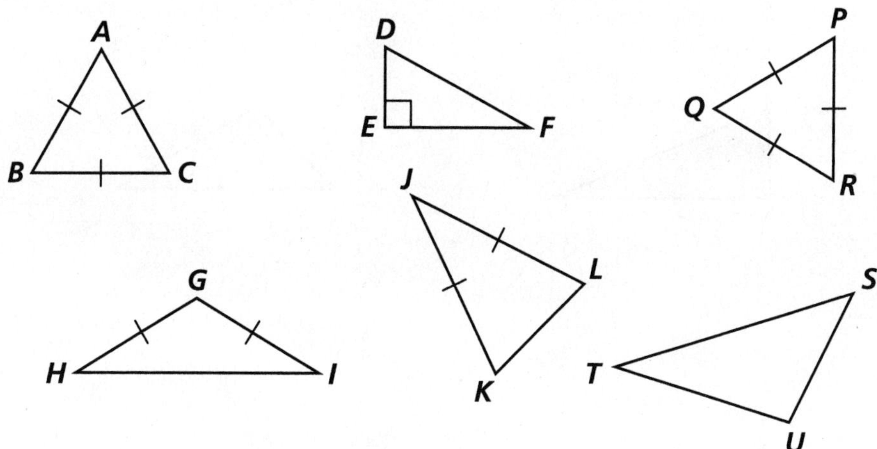

17. Write the letter names of the scalene triangles.

18. Write the letter names of the equilateral triangles.

19. Write the letter names of the isosceles triangles.

Show your work on your paper or in your journal.

► Identify Sides of a Triangle

Name each triangle by its sides. Explain your thinking.

20. 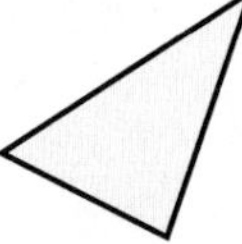

21. 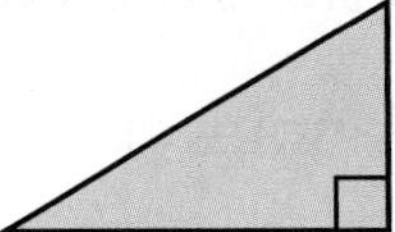

22. 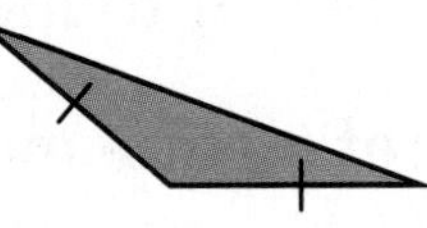

23. 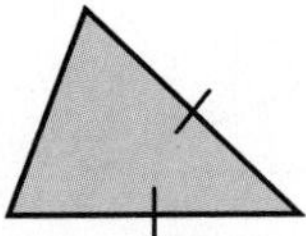

24. 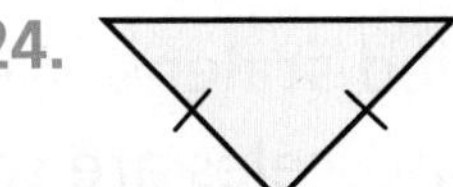

25. 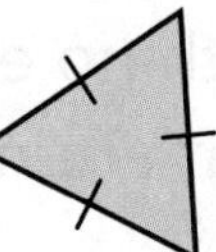

26. 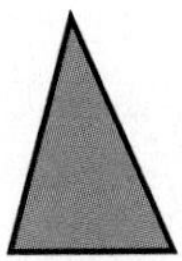

27. 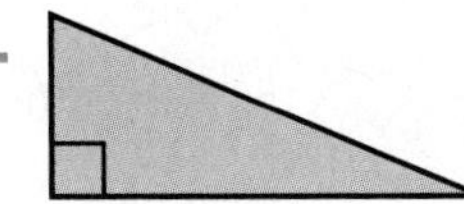

28. 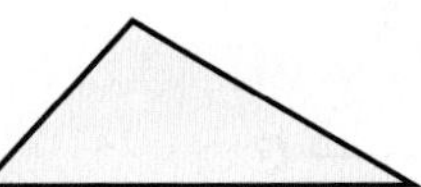

29. 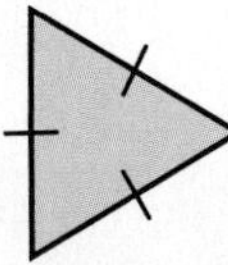

30. 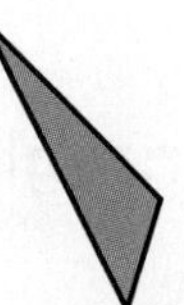

31. 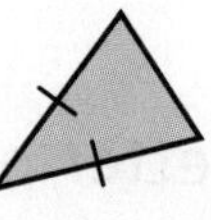

32. Explain how sides make triangles different from each other.

Use Activity Workbook page 103.

► Sort Triangles in Different Ways

33. Write a capital letter and a lowercase letter inside each triangle below, using the keys at the right.

Cut out the triangles and use the diagram below to sort them in different ways.

acute = a
obtuse = o
right = r

Isosceles = I
Scalene = S
Equilateral = E

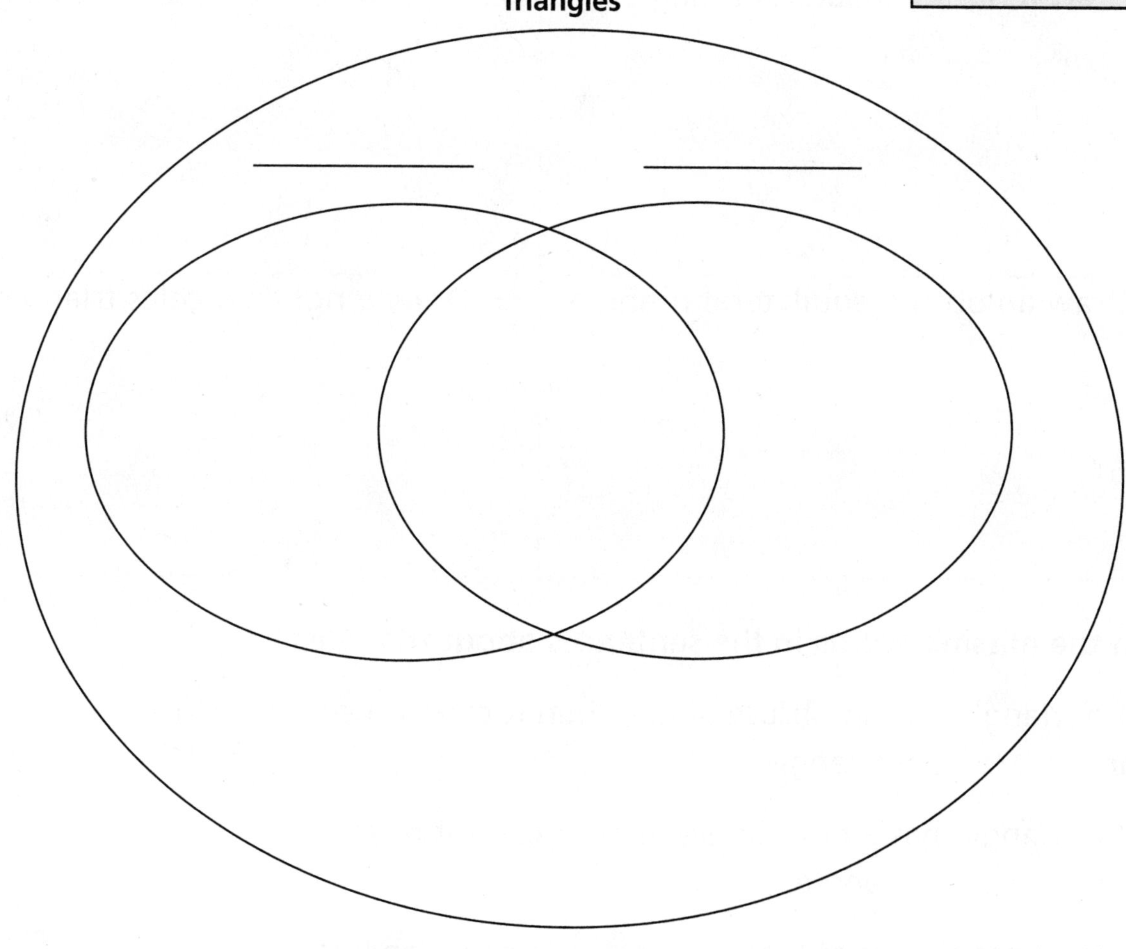

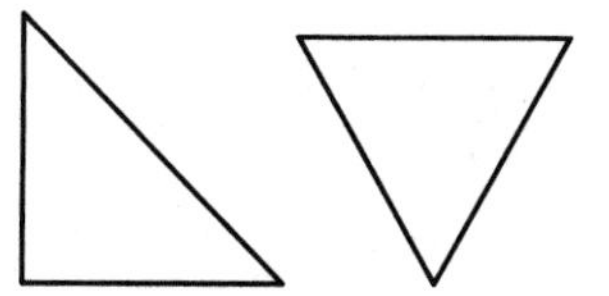
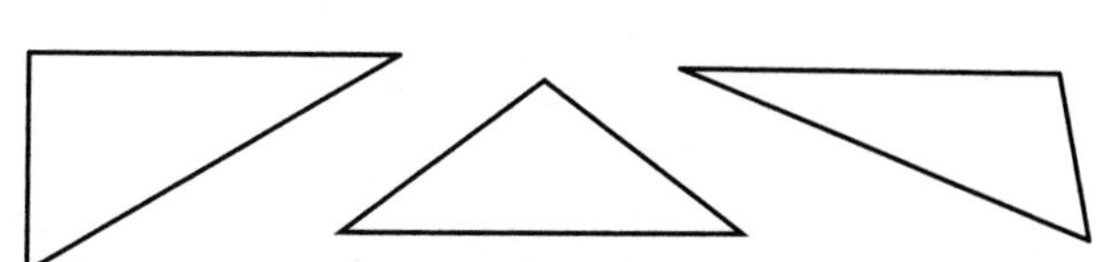
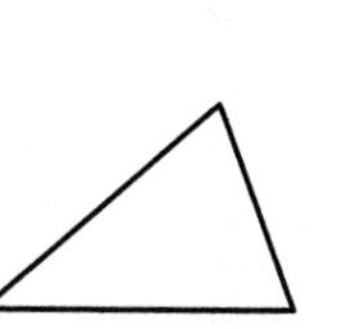
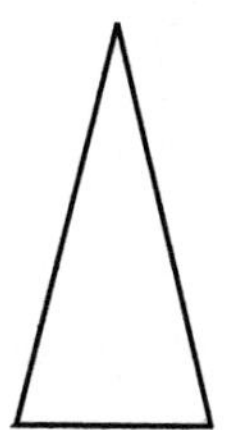

Use Activity Workbook page 104.

▶ Possible Ways to Name Triangles

Draw each triangle. If you can't, explain why.

34. Draw a right scalene triangle.	35. Draw an obtuse scalene triangle.
36. Draw a right equilateral triangle.	37. Draw an acute isosceles triangle.
38. Draw an obtuse equilateral triangle.	39. Draw a right isosceles triangle.

Fill in the missing words in the sentences about triangles.

40. If a triangle has an obtuse angle, then it cannot be an ________ triangle.

41. If a triangle has a right angle, then it cannot have an ________ angle.

42. Every triangle has at least ________ acute angles.

▶ Add Angle Measures

Two angles can be put together to form another angle. The measure of the whole angle is the sum of the measures of the smaller angles. The measure of the whole angle shown is 105°.

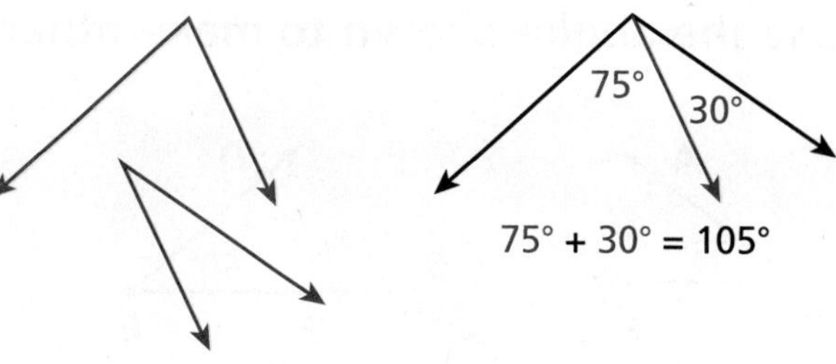

What kind of angle is formed when the two angles are put together? What is its measure?

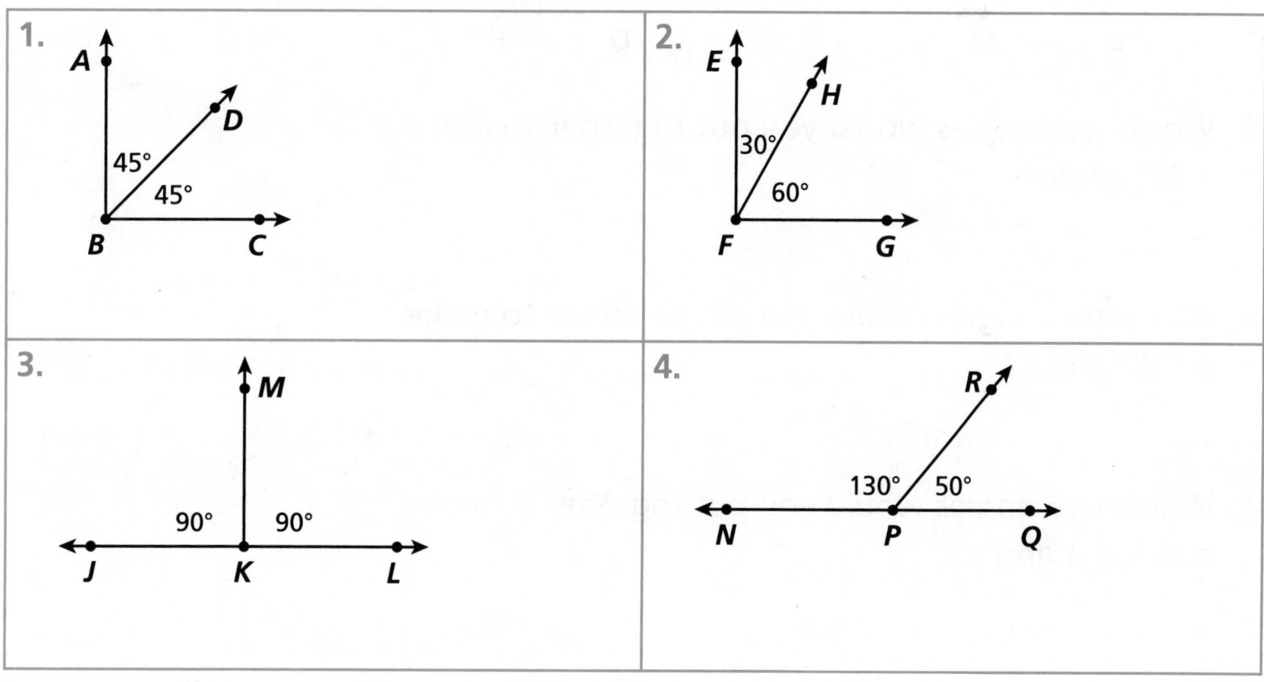

5. An angle is made from two angles with measures 80° and 70°. Write and solve an equation to find the measure of the whole angle.

▶ Put Angles Together

Use the angles shown to make other angles.

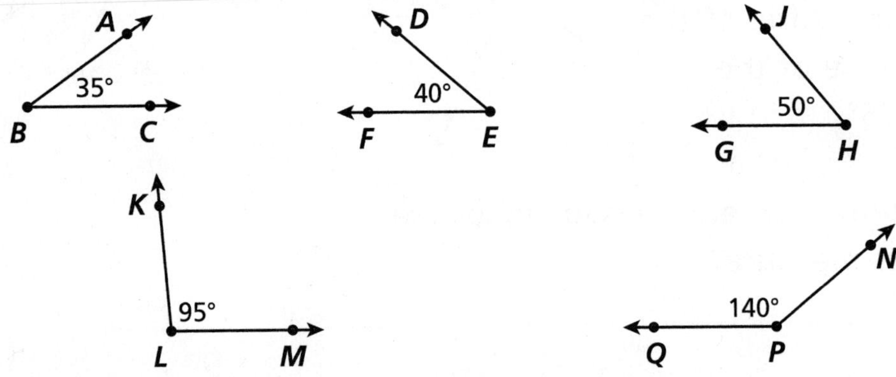

6. Which two angles would you put together to make a 75° angle?

7. Which two angles would you put together to make a 145° angle?

8. Which two angles would you put together to make a straight angle?

9. Which two angles would you put together to make a right angle?

10. If you put all five angles together, what would be the measure of the whole angle? What kind of figure would you form?

11. Use a protractor and straightedge to draw the angle formed by putting $\angle ABC$ and $\angle KLM$ together. Show its measure.

Show your work on your paper or in your journal.

► Subtract Angle Measures

Write an equation to find the unknown angle measure.

12. 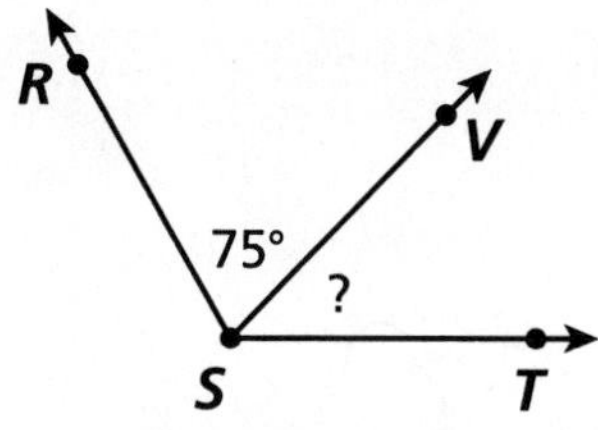

The measure of ∠*RST* is 120°. What is the measure of ∠*VST*?

13. 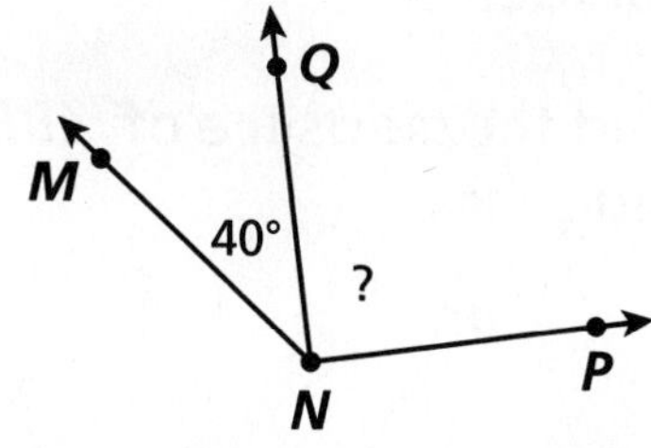

The measure of ∠*MNP* is 130°. What is the measure of ∠*QNP*?

14. 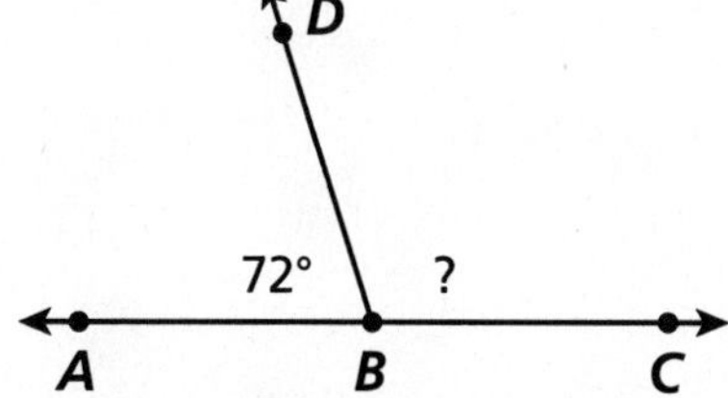

The measure of ∠*ABC* is 180°. What is the measure of ∠*DBC*?

15. 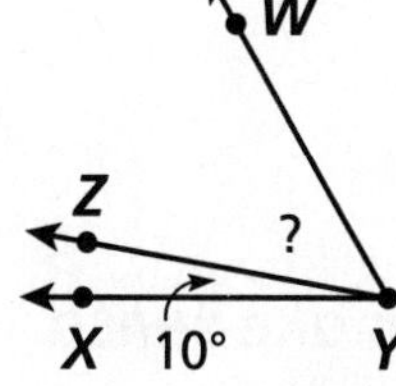

The measure of ∠*XYW* is 60°. What is the measure of ∠*ZYW*?

16. Draw your own angle problem and share it with a partner.

17. When a right angle is made from two smaller angles, what must be true about the smaller angles?

18. When a straight angle is made from two smaller angles, what must be true about the smaller angles?

► What's the Error?

Dear Math Students,

I want to find the measure of $\angle DBE$ in the following diagram.

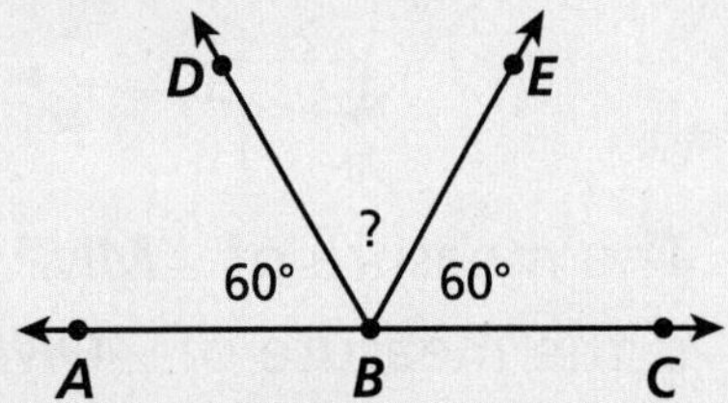

I wrote and solved this equation.

$$180^\circ - (60^\circ + 60^\circ) = x$$
$$180^\circ - 60^\circ + 60^\circ = x$$
$$120^\circ + 60^\circ = x$$
$$180^\circ = x$$

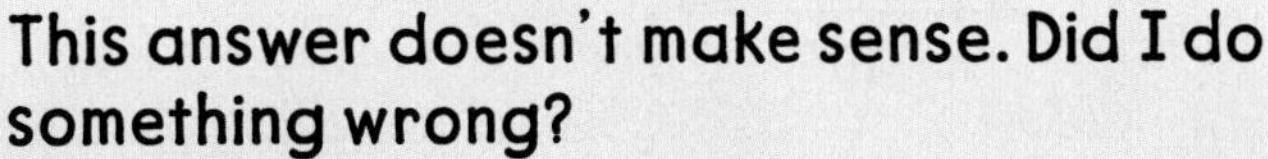

This answer doesn't make sense. Did I do something wrong?

Your friend,
Puzzled Penguin

19. Write a response to Puzzled Penguin.

Show your work on your paper or in your journal.

▶ Add Angle Measures

Use an equation to solve.

1. The ski jumper shown makes angles with her skis as shown. What is the sum of the angles?

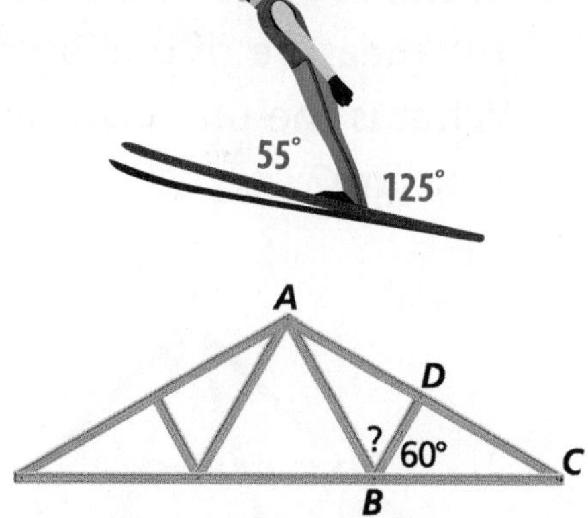

2. In the roof framework shown, ∠*ABD* and ∠*DBC* have the same measure. What is the measure of ∠*DBA*? What is the measure of ∠*ABC*?

3. In the simple bridge structure shown, the measure of ∠*RSV* is 30° and ∠*VST* is a right angle. What is the measure of ∠*RST*?

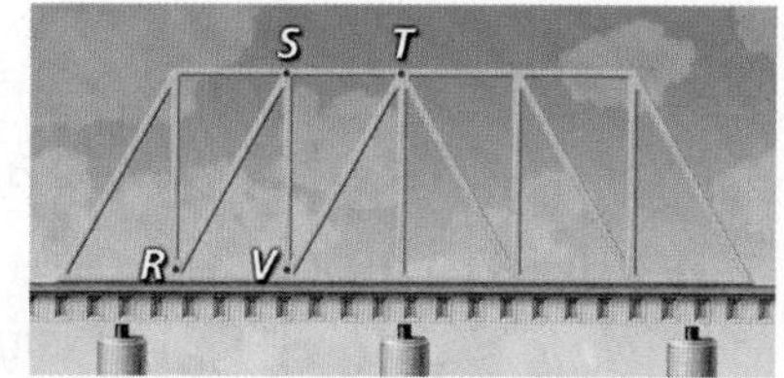

The circle at the right represents all of the students in a class. Each section represents the students in the class who chose a certain type of animal as their favorite type of pet. The angle measures for some sections are given.

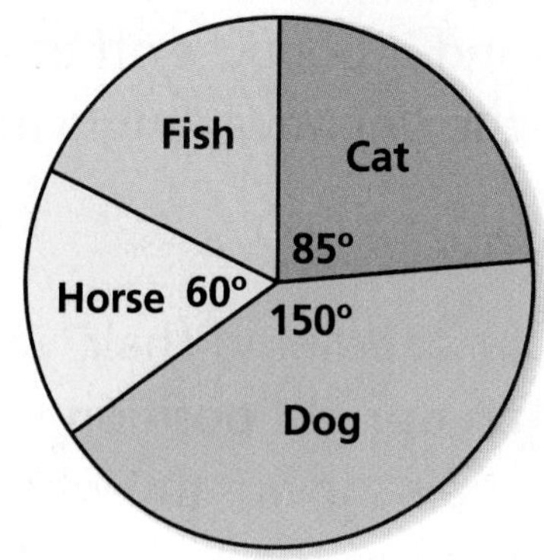

4. What is the sum of the angle measures for Cat, Dog, and Horse?

5. What is the total angle measure for the circle? What is the angle measure for Fish?

▶ Subtract Angle Measures

Use an equation to solve.

6. In the roof framework shown, the measure of one angle is 80°. What is the unknown angle measure?

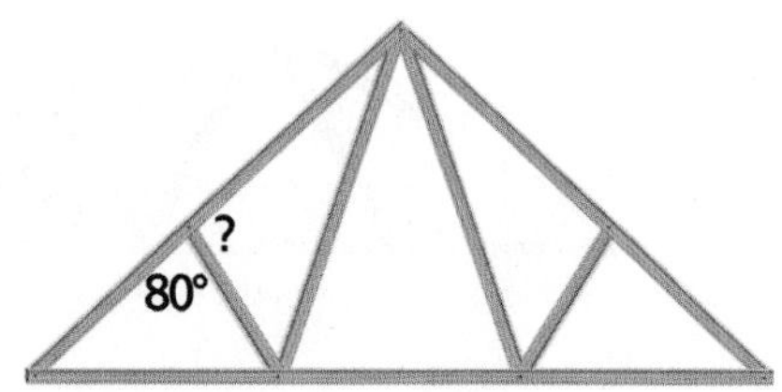

7. The railing on a stairway makes a 50° angle with the upright post. What is the unknown angle measure in the diagram?

8. When different items are poured, they form a pile in the shape of a cone. The diagram shows a pile of sand. What is the angle the sand makes with the ground?

9. In a miniature golf game, a player hits a ball against a wall at an angle with measure 35° and it bounces off at an angle of 20°. What is the unknown angle measure in the diagram?

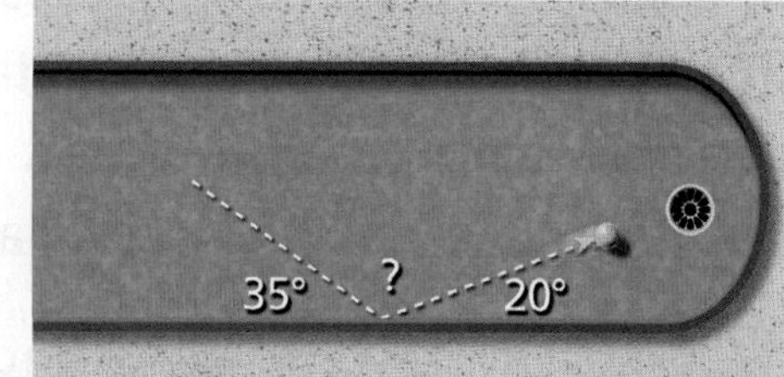

10. In a reclining chair, you can push back from an upright position to sit at an angle. In the chair shown, the whole angle between the back of the chair and the seat of the chair is 130°. Find the unknown angle measure to find by how much the chair is reclined from upright.

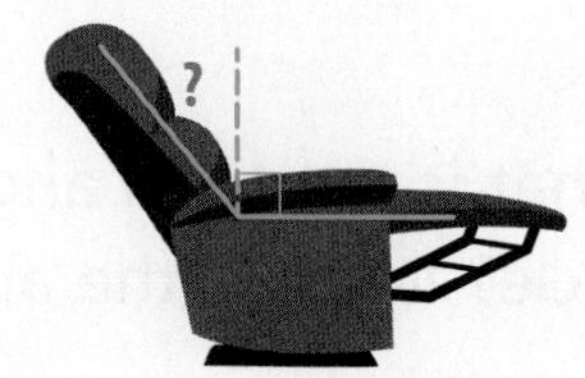

Dear Family,

Your child has been learning about geometry throughout this unit. In this second half of the unit, your child will be learning how to recognize and describe a group of geometric figures called quadrilaterals, which get their name because they have four (*quad-*) sides (*-lateral*). Five different kinds of quadrilaterals are shown here.

Share with your family the Family Letter on Activity Workbook page 105.

Square
4 equal sides
opposite sides parallel
4 right angles

Rectangle
2 pairs of parallel sides
4 right angles

Rhombus
4 equal sides
opposite sides parallel

Parallelogram
2 pairs of parallel sides

Trapezoid
exactly 1 pair of opposite sides parallel

If you have any questions or comments, please call or write to me.

Sincerely,
Your child's teacher

This unit includes the Common Core Standards for Mathematical Content for Operations and Algebraic Thinking, 4.OA.5; Geometry, 4.G.1, 4.G.2, 4.G.3; and all the Mathematical Practices.

Estimada familia:

Durante esta unidad, su niño ha estado aprendiendo acerca de geometría. En esta parte de la unidad, su niño aprenderá cómo reconocer y describir un grupo de figuras geométricas llamadas cuadriláteros, que reciben ese nombre porque tienen cuatro *(quadri-)* lados *(-lateris)*. Aquí se muestran cinco tipos de cuadriláteros:

Cuadrado
4 lados iguales
lados opuestos paralelos
4 ángulos rectos

Rectángulo
2 pares de lados paralelos
4 ángulos rectos

Muestra a tu familia la Carta a la familia de la página 106 del Cuaderno de actividades y trabajo.

Rombo
4 lados iguales
lados opuestos paralelos

Paralelogramo
2 pares de lados paralelos

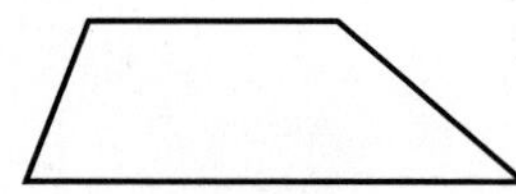

Trapecio
exactamente 1 par de lados paralelos opuestos

Si tiene alguna pregunta o algún comentario, por favor comuníquese conmigo.

Atentamente,
El maestro de su niño

Esta unidad incluye los Common Core Standards for Mathematical Content for Operations and Algebraic Thinking, 4.OA.5; Geometry, 4.G.1, 4.G.2, 4.G.3; and all the Mathematical Practices.

8-7
Class Activity

Use Activity Workbook page 107.

VOCABULARY
parallel

► Define Parallel Lines

The lines or line segments in these pairs are **parallel**.

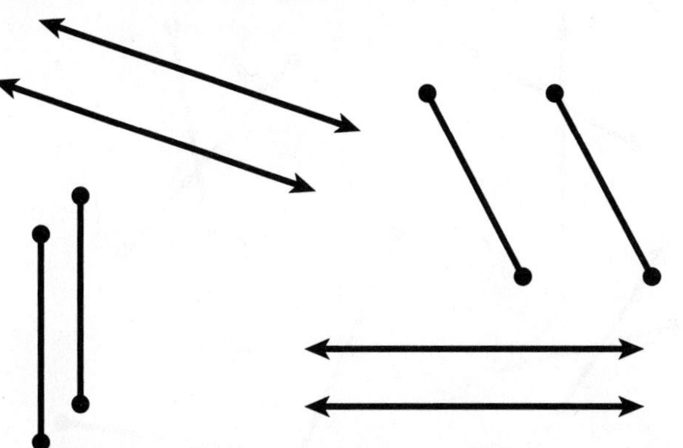

The lines or line segments in these pairs are not parallel.

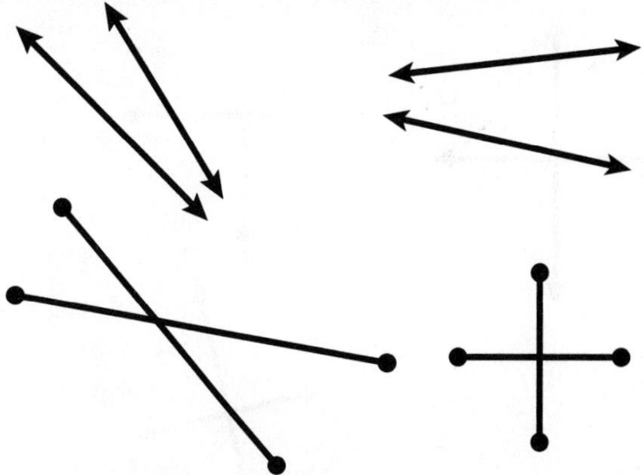

1. What do you think it means for two lines to be parallel?

► Draw Parallel Lines

2. Draw and label a pair of parallel lines.

3. Draw and label a figure with one pair of parallel line segments.

Use Activity Workbook page 108.

VOCABULARY
perpendicular

▶ Define Perpendicular Lines

The lines or line segments in these pairs are **perpendicular**.

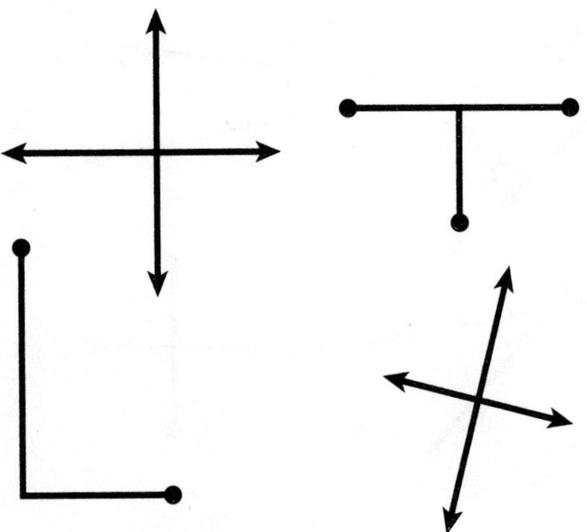

The lines or line segments in these pairs are not perpendicular.

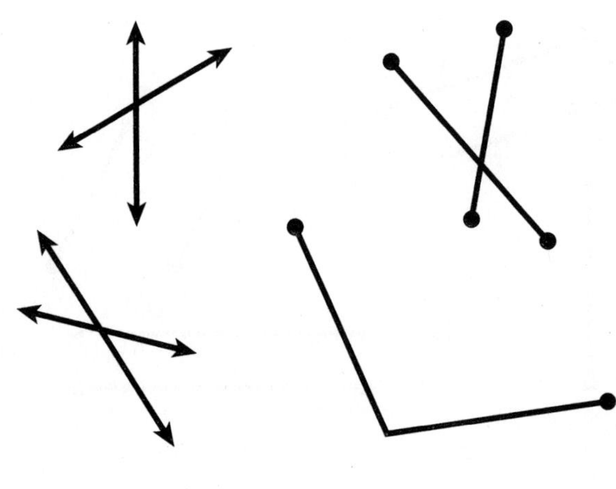

4. What do you think it means for two lines to be perpendicular?

▶ Draw Perpendicular Lines

5. Draw and label a pair of perpendicular lines.

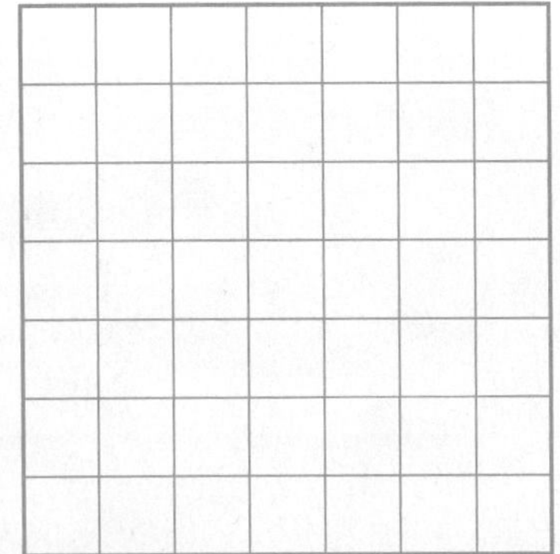

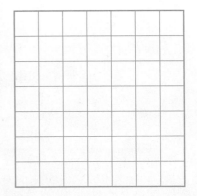

6. Draw and label a figure with one pair of perpendicular line segments.

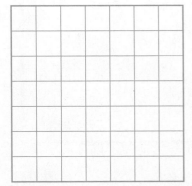

Show your work on your paper or in your journal.

► Identify Types of Lines

Tell whether each pair of lines is parallel, perpendicular, or neither.

7. 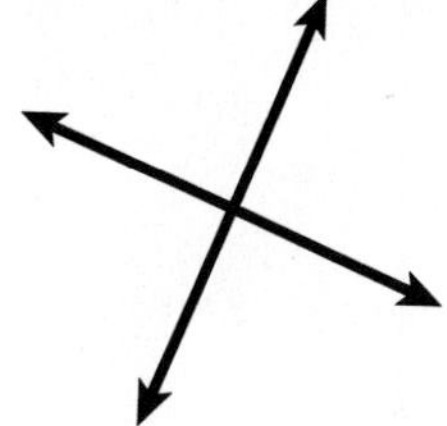

8. 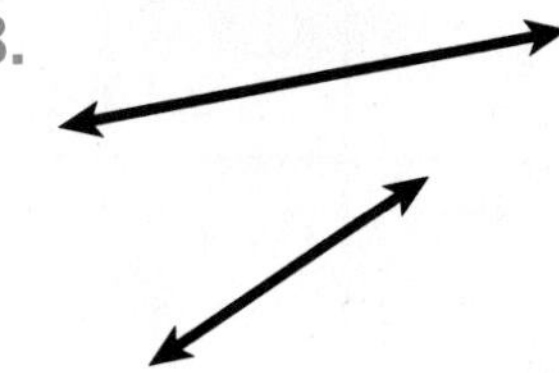

9. 

10. 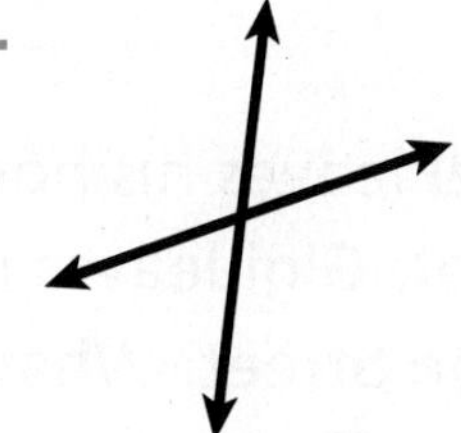

11. Draw a pair of parallel line segments.

12. First, draw a line segment 3 cm long. Then, draw a line segment 6 cm long that looks perpendicular to your first line segment.

► Lines on a Map

Use the map.

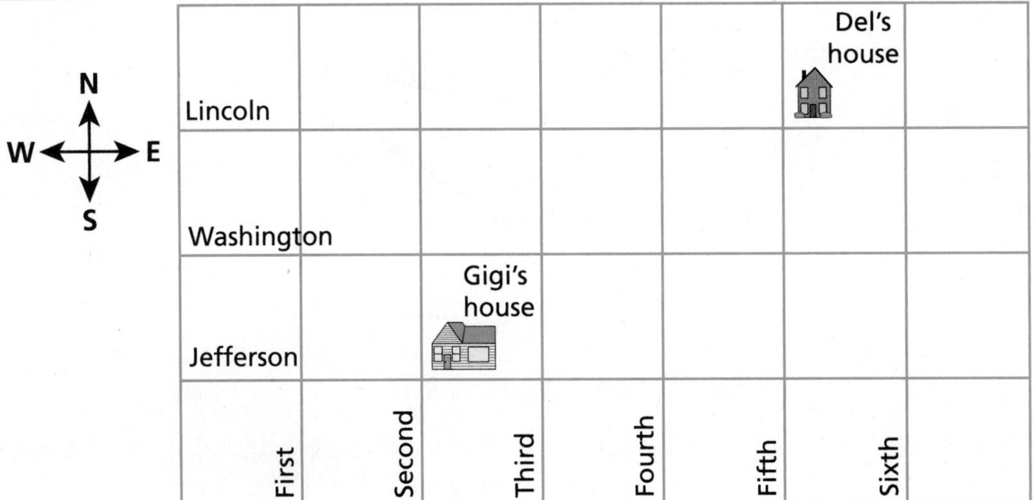

13. On Wednesday, Del leaves his house and walks West along Lincoln Street. Gigi leaves her house and walks East along Jefferson Street. What kind of lines are Lincoln Street and Jefferson Street?

14. Will Del and Gigi ever meet? If so, where?

15. On Friday, Del leaves his house and walks South along Fifth Street. Gigi leaves her house and walks East along Jefferson Street. What kind of lines are Fifth Street and Jefferson Street?

16. Will Del and Gigi ever meet? If so, where?

VOCABULARY
quadrilateral
adjacent
opposite

▶ Identify Sides of Quadrilaterals

Look at these quadrilaterals.

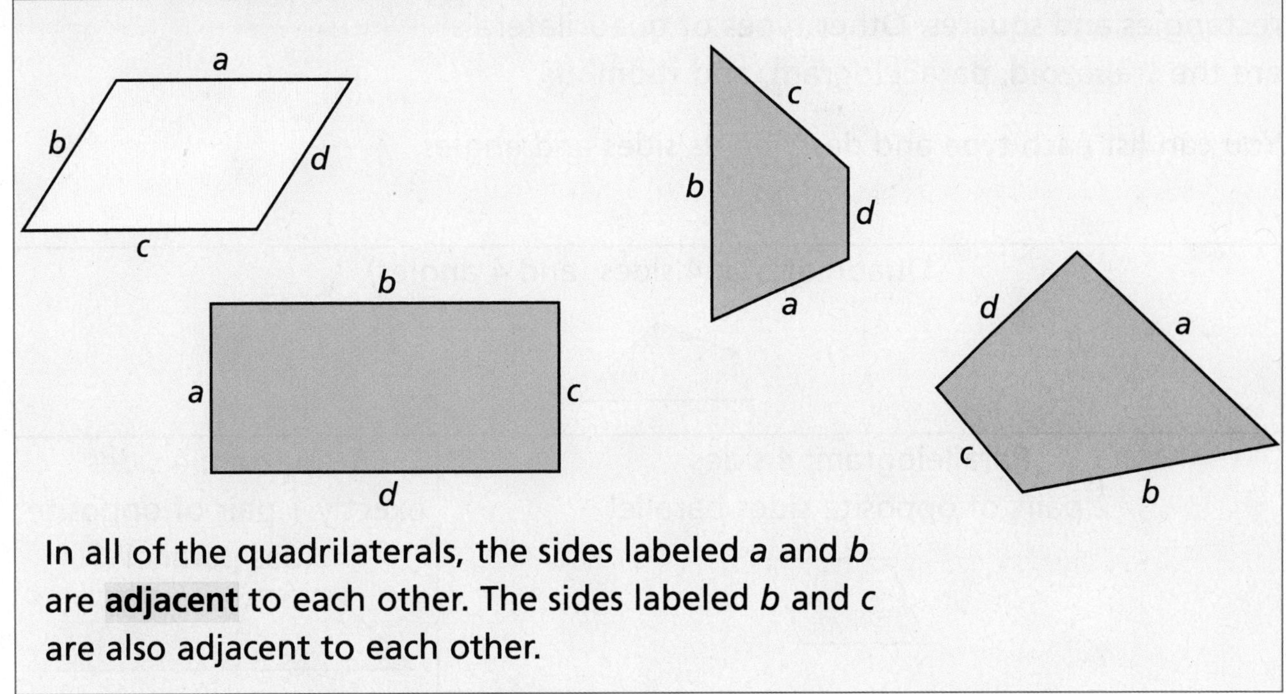

In all of the quadrilaterals, the sides labeled *a* and *b* are **adjacent** to each other. The sides labeled *b* and *c* are also adjacent to each other.

1. What do you think it means for two sides to be adjacent?

2. Which other sides are adjacent to each other?

In all of the quadrilaterals, the sides labeled *a* and *c* are **opposite** each other.

3. What do you think it means for two sides to be opposite each other?

4. Which other sides are opposite each other?

VOCABULARY
trapezoid
parallelogram
rhombus

► Identify Types of Quadrilaterals

Some quadrilaterals are special because they have parallel sides or right angles. You already know about rectangles and squares. Other types of quadrilaterals are the **trapezoid**, **parallelogram**, and **rhombus**.

You can list each type and describe its sides and angles.

Quadrilateral: 4 sides (and 4 angles)	
Parallelogram: 4 sides 2 pairs of opposite sides parallel Rhombus: 4 sides 2 pairs of opposite sides parallel 4 equal sides Rectangle: 4 sides 2 pairs of opposite sides parallel 4 right angles Square: 4 sides 2 pairs of opposite sides parallel 4 right angles 4 equal sides	Trapezoid: 4 sides exactly 1 pair of opposite sides parallel

Use Activity
Workbook page 109.

► Draw Special Quadrilaterals

5. Draw a quadrilateral that has exactly one pair of opposite sides parallel. What type of quadrilateral is it?

6. Draw a quadrilateral that has two pairs of opposite sides parallel. What type of quadrilateral is it? Is there more than one answer?

7. Draw a quadrilateral that has two pairs of opposite sides parallel, 4 equal sides, and no right angles. What type of quadrilateral is it?

8-8
Class Activity

Use Activity Workbook page 110.

► Identify Relationships

Why is each statement below true?

8. A rhombus is always a parallelogram, but a parallelogram isn't always a rhombus.

9. A rectangle is a parallelogram, but a parallelogram is not necessarily a rectangle.

10. A square is a rectangle, but a rectangle does not have to be a square.

11. Complete the category diagram by placing each word in the best location.

    Quadrilateral
    Trapezoid
    Parallelogram
    Rectangle
    Rhombus
    Square

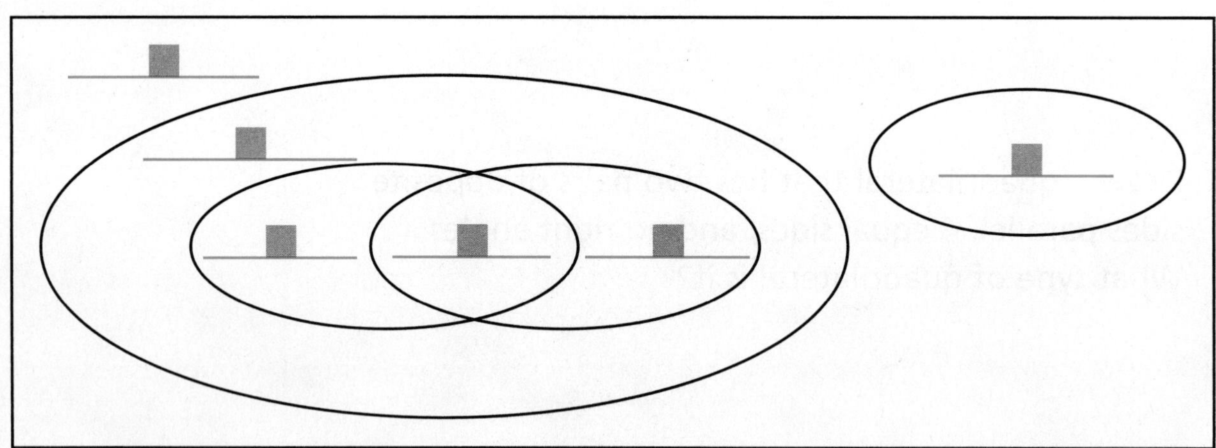

8-9
Class Activity

VOCABULARY
diagonal

▶ Use Diagonals to Make Triangles

A **diagonal** connects opposite angles of a quadrilateral. You can make triangles by drawing a diagonal on a quadrilateral.

List all names for each quadrilateral in Exercises 1–3. Then use letters to name the triangles you can make with the diagonals and tell what kind of triangles they are.

1. 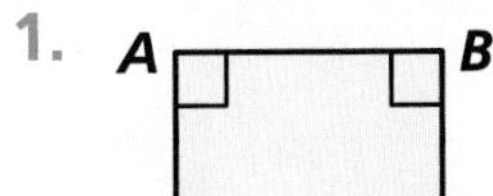

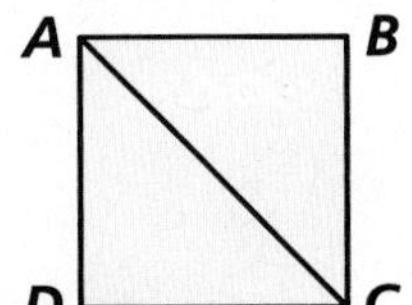

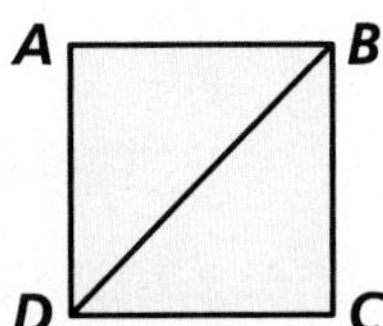

2. 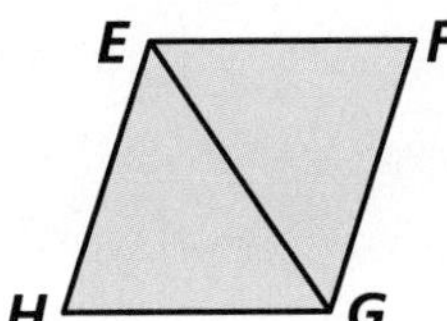

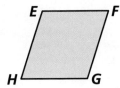

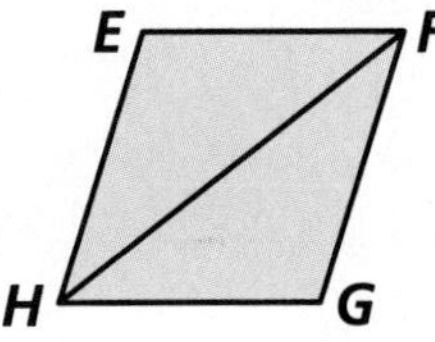

3. 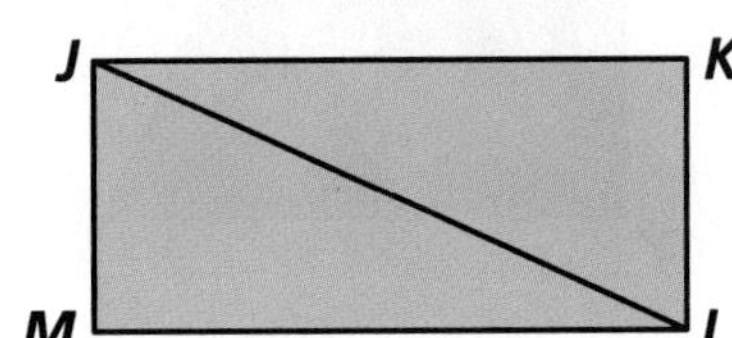

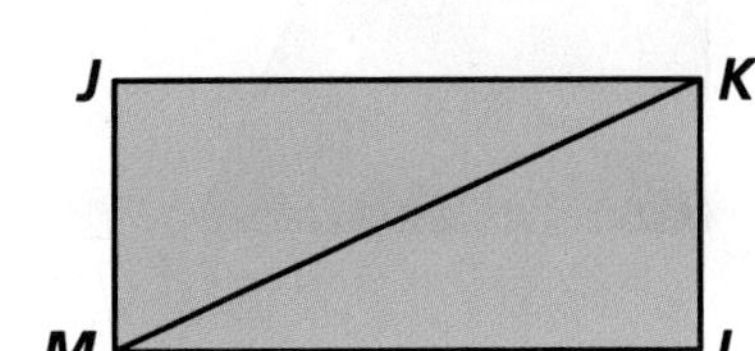

▶ Use Diagonals to Make Triangles (continued)

List all names for each quadrilateral in Exercises 4–6. Then use letters to name the triangles you can make with the diagonals and tell what kind of triangles they are.

4.

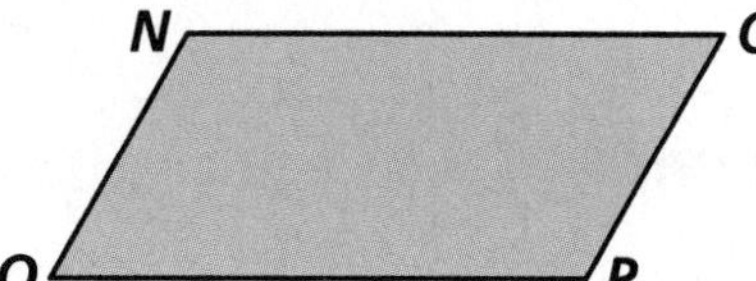

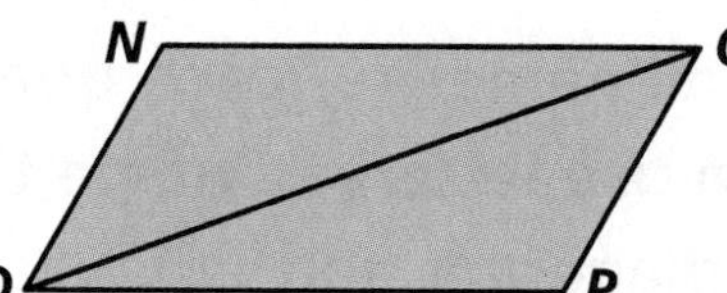

5.

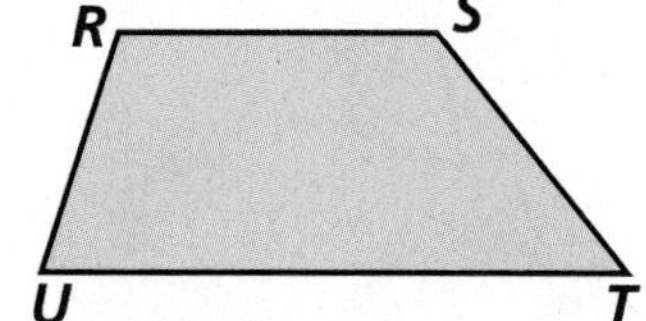

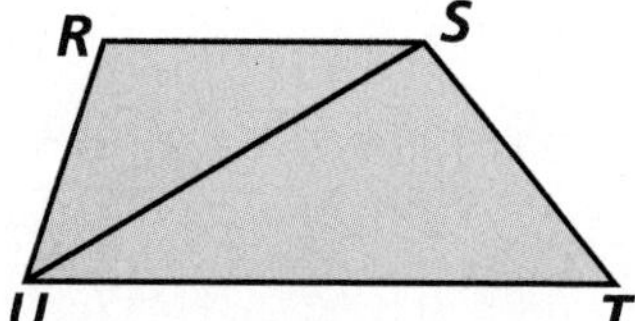

6.

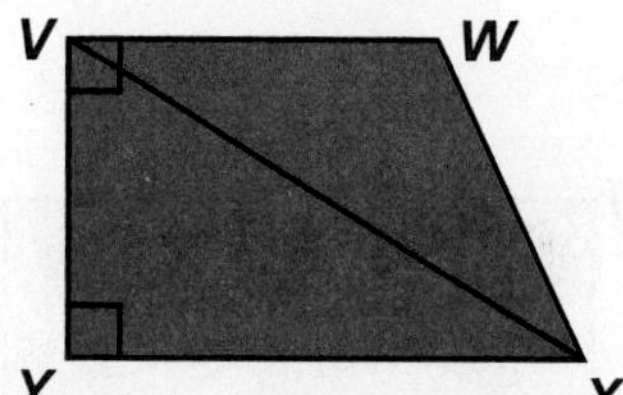

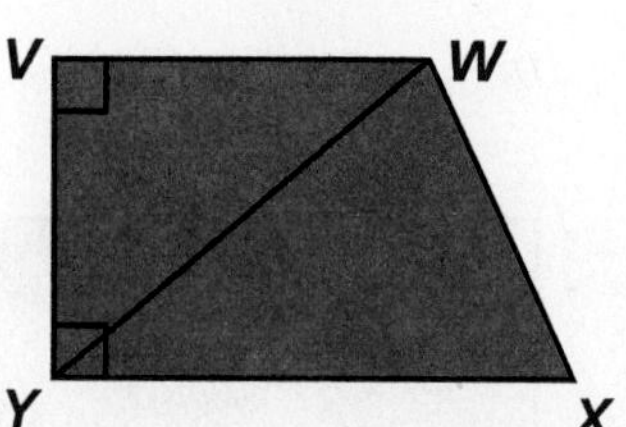

Use Activity Workbook page 113.

► Build Quadrilaterals With Triangles

You can make a quadrilateral by joining the equal sides of two triangles that are the same size and shape.

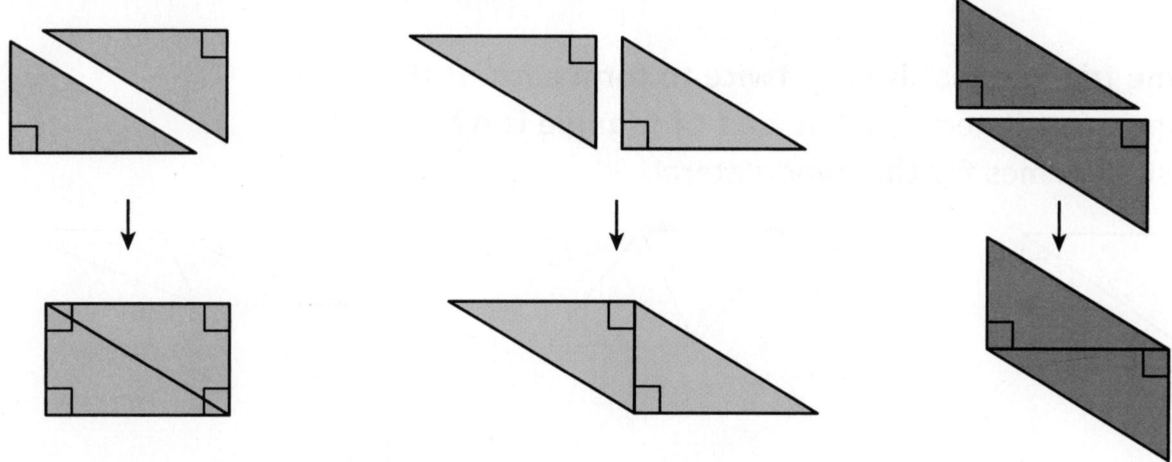

Cut out the triangles below. For each exercise, glue two of the triangles on this paper so that the stated sides are joined. Then write the name of the quadrilateral.

7. $\overline{AB}$ is joined to $\overline{AB}$

8. $\overline{AC}$ is joined to $\overline{AC}$

9. $\overline{BC}$ is joined to $\overline{BC}$

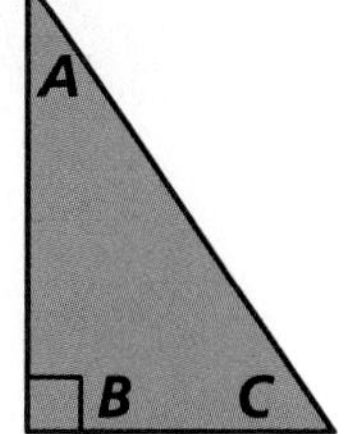

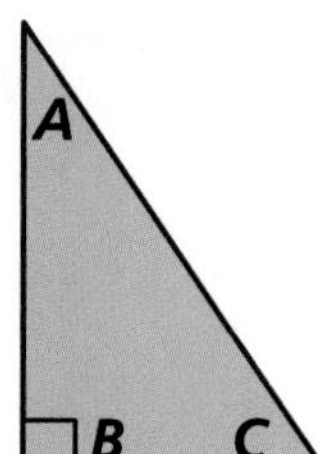

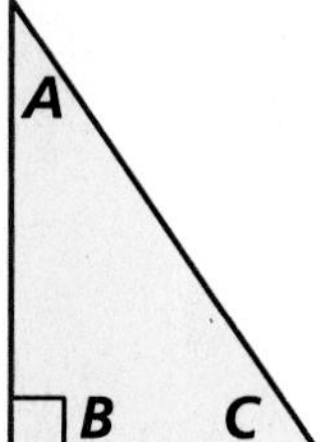

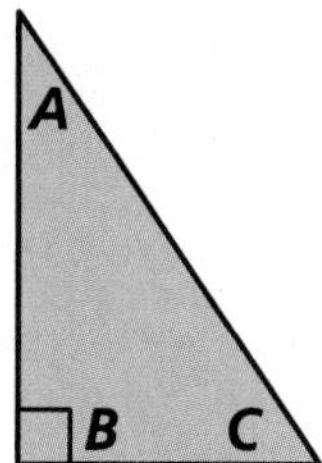

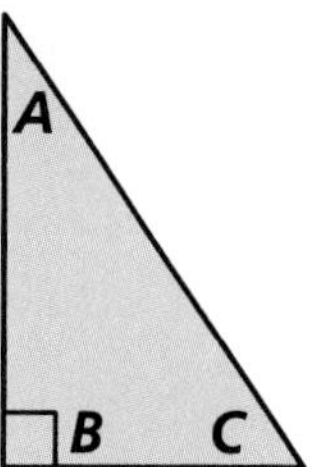

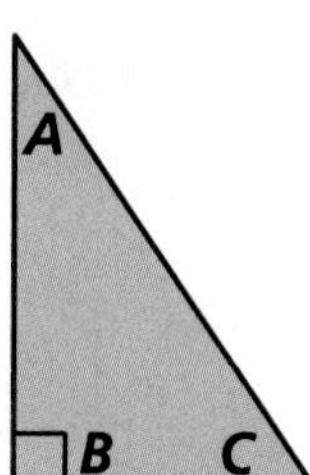

Use Activity Workbook page 114.

► Match Quadrilaterals with Triangles

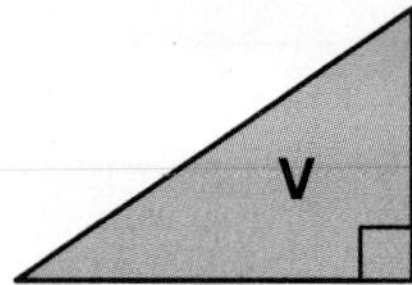

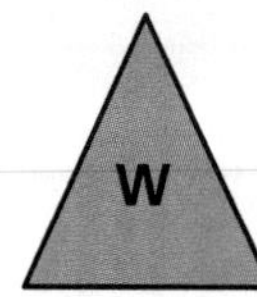

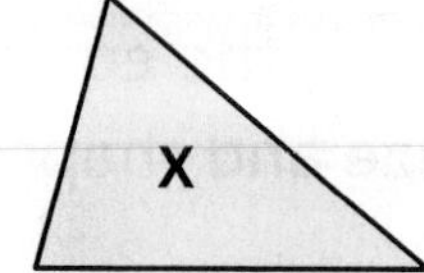

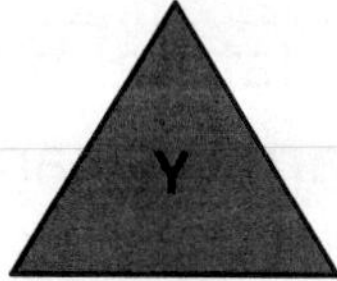

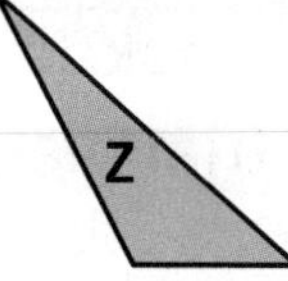

Name the triangle that is used twice to form each of the following quadrilaterals. What kind of triangle is it? Then list all names for the quadrilateral.

10. 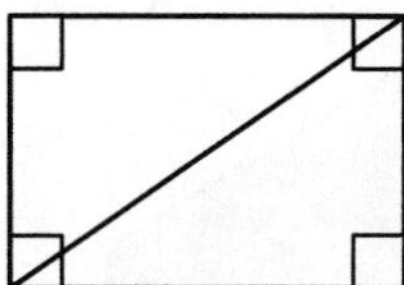

11. 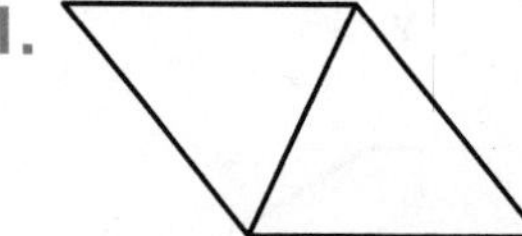

12. 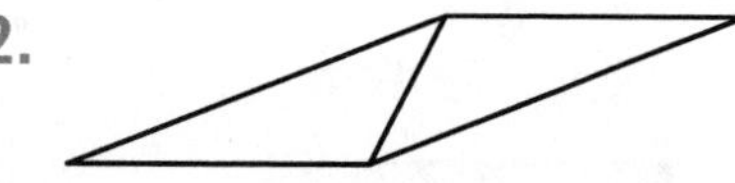

13. 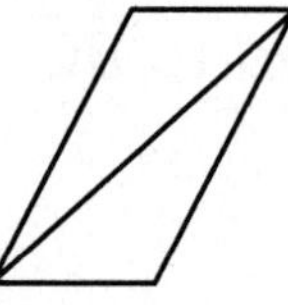

14. 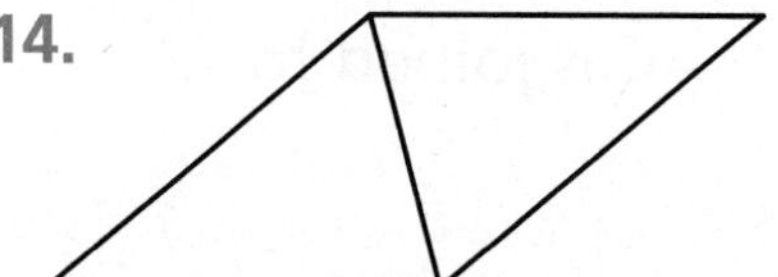

15. 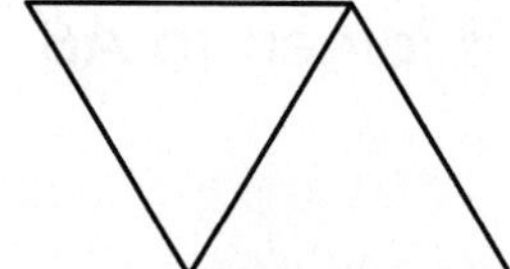

VOCABULARY
vertex

▶ Draw Perpendicular Lines in Triangles

A **vertex** is a point shared by two sides of a polygon.

Draw a scalene triangle *ABC*.

16. What is true about the sides of your triangle?

17. Draw a segment from one vertex so that it is perpendicular to the opposite side. Label the segment and mark the right angles.

18. Name the triangles formed. What kind of triangles are they?

19. Are the triangles you formed the same size and shape?

Draw an isosceles triangle *JKL* for Exercises 20–23.

20. What is true about the sides of your triangle?

21. Draw a segment from the vertex between the equal sides of the triangle so that it is perpendicular to the opposite side. Label the segment and mark the right angles.

► Draw Perpendicular Lines in Triangles (continued)

Use your isosceles triangle *JKL* to answer the questions.

22. Name the segments formed by the perpendicular segment in $\triangle JKL$. What is true about the lengths of the segments?

23. Name the triangles formed. Are they the same size and shape?

► What's the Error?

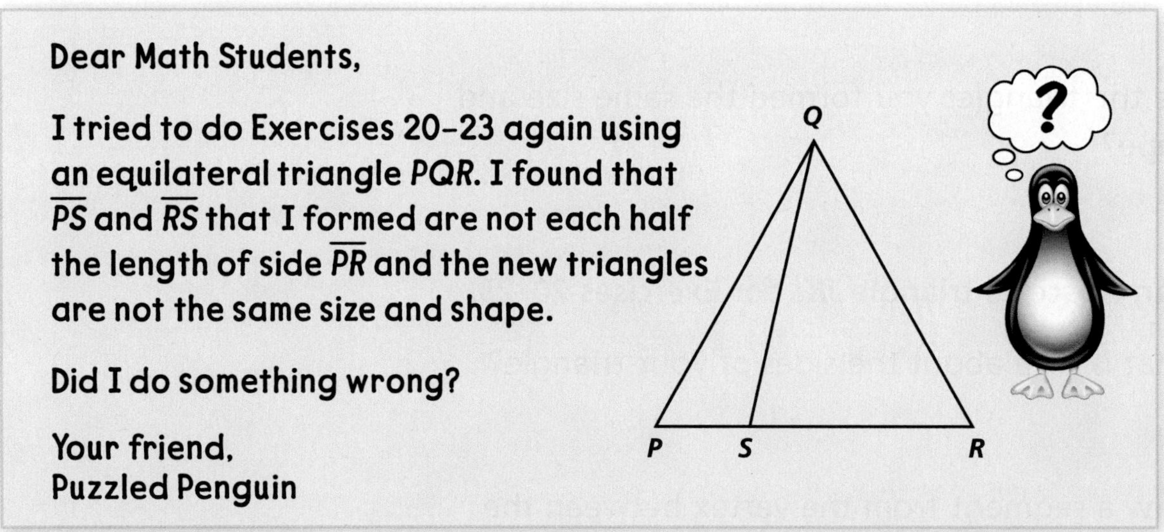

Dear Math Students,
I tried to do Exercises 20–23 again using an equilateral triangle *PQR*. I found that $\overline{PS}$ and $\overline{RS}$ that I formed are not each half the length of side $\overline{PR}$ and the new triangles are not the same size and shape.
Did I do something wrong?
Your friend,
Puzzled Penguin
Q
P
S
R
?

24. Write a response to Puzzled Penguin.

VOCABULARY
polygon

► Sort Polygons by Angles

Triangles and quadrilaterals are examples of **polygons**.

Use these polygons for Exercises 1–5.

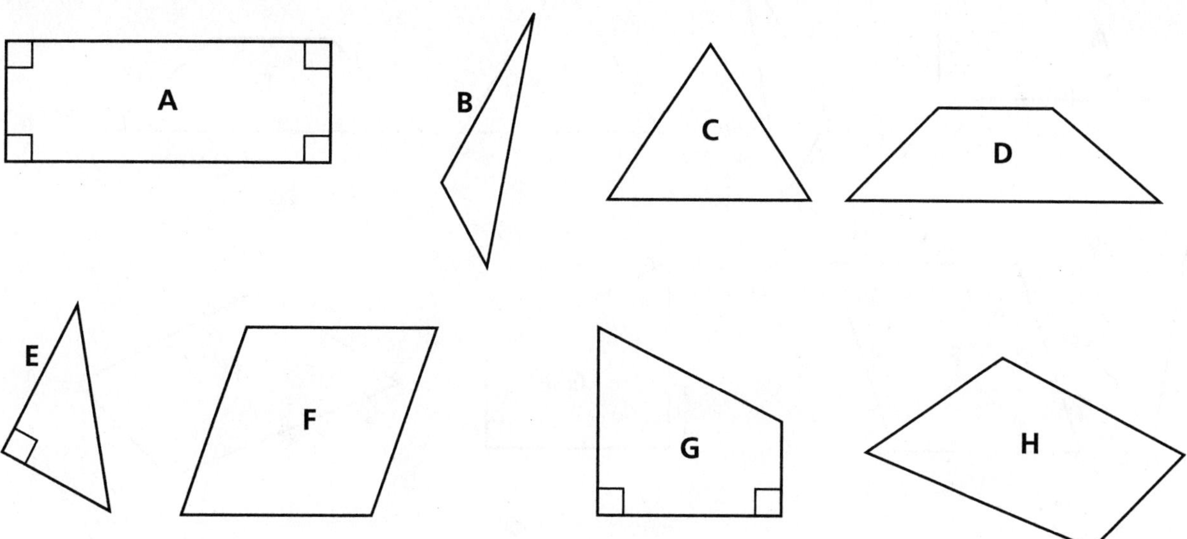

1. Which figures have one or more acute angles?

2. Which figures have one or more right angles?

3. Which figures have one or more obtuse angles?

4. Which figures have both acute angles and right angles?

5. Which figures have both acute angles and obtuse angles?

▶ Sort Polygons by Sides

Use these polygons for Exercises 6–10.

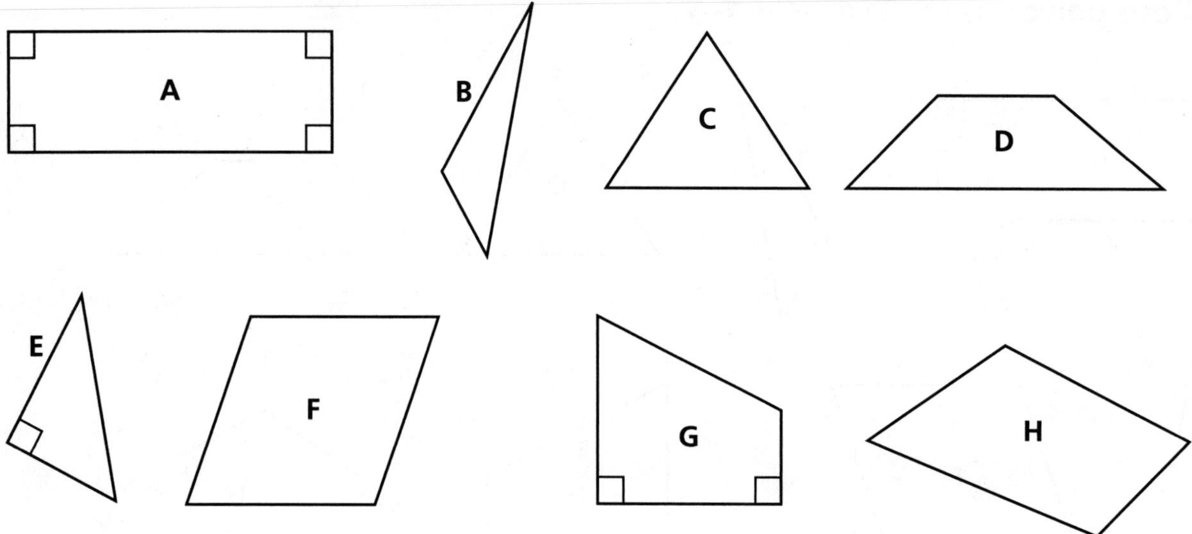

6. Which figures have perpendicular sides?

7. Which figures have exactly one pair of opposite sides parallel?

8. Which figures have two pairs of opposite sides parallel?

9. Which figures have both parallel and perpendicular sides?

10. Which figures have no parallel or perpendicular sides?

VOCABULARY
line symmetry
line of symmetry

▶ Identify Line Symmetry in Figures

A plane figure has **line symmetry** if it can be folded along a line so the two halves match exactly. The fold is called a **line of symmetry**.

Does the figure have line symmetry? Write *yes* or *no*.

1. 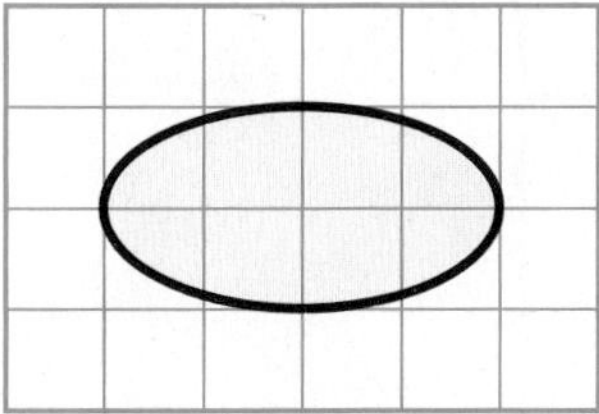

2. 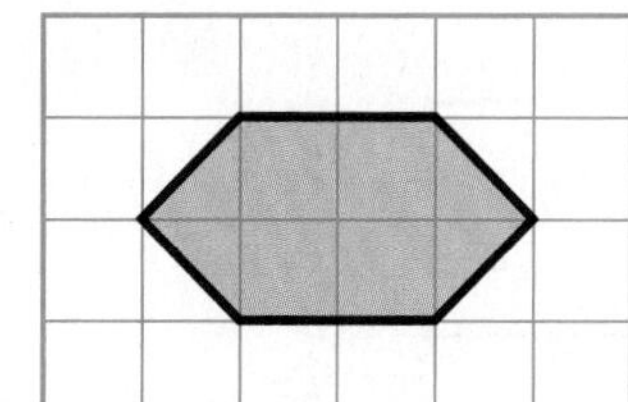

3. 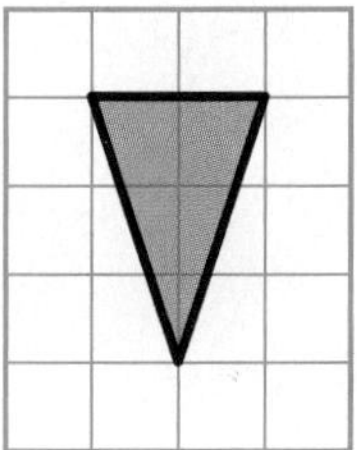

4. 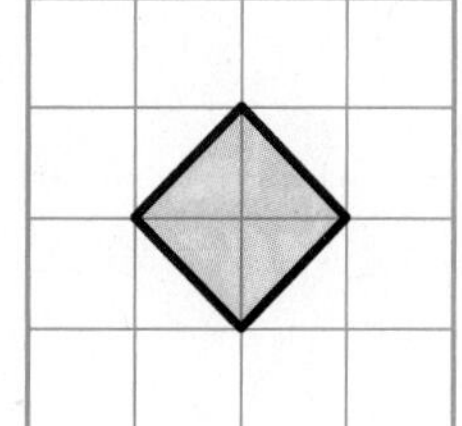

5. 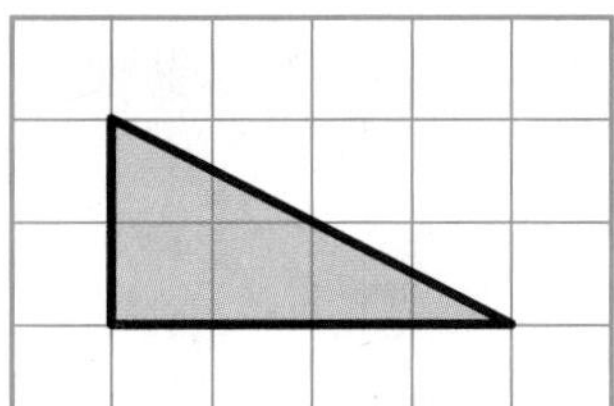

6. 

Use Activity Workbook page 117.

▶ Draw Lines of Symmetry

A line of symmetry divides a figure or design into two matching parts.

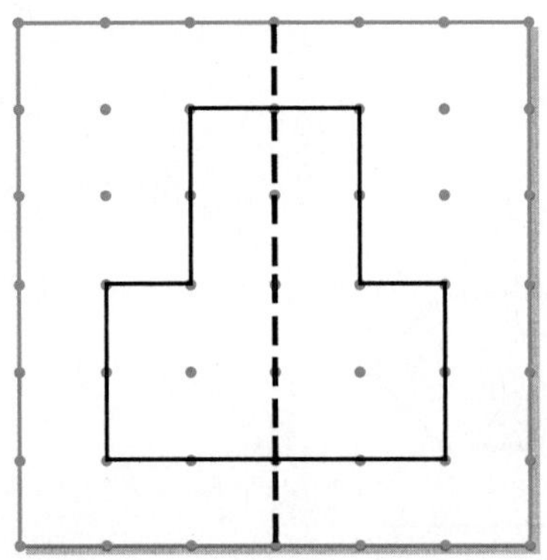

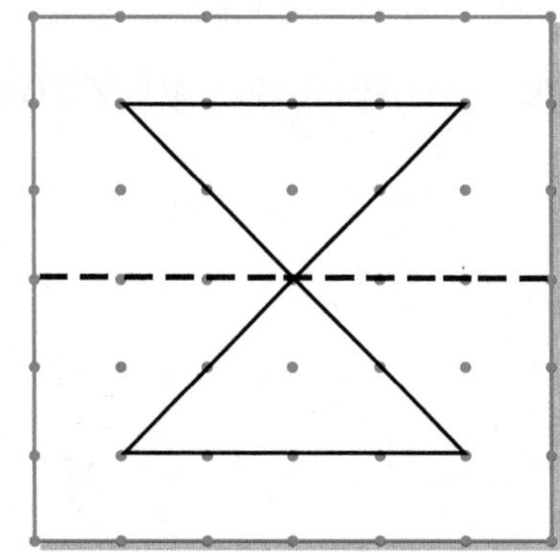

Draw the line of symmetry in the figure or design.

7. 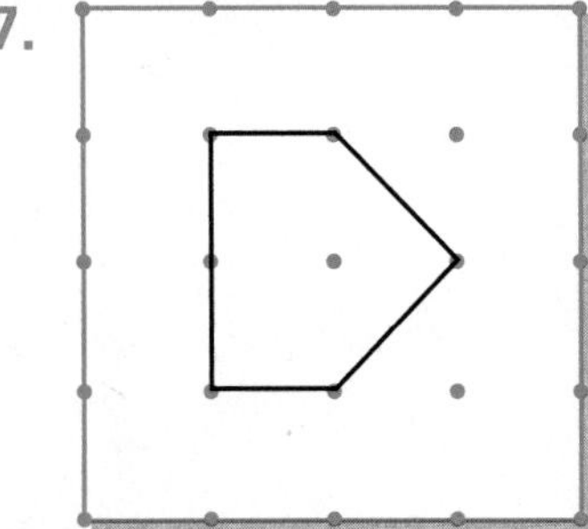

8. 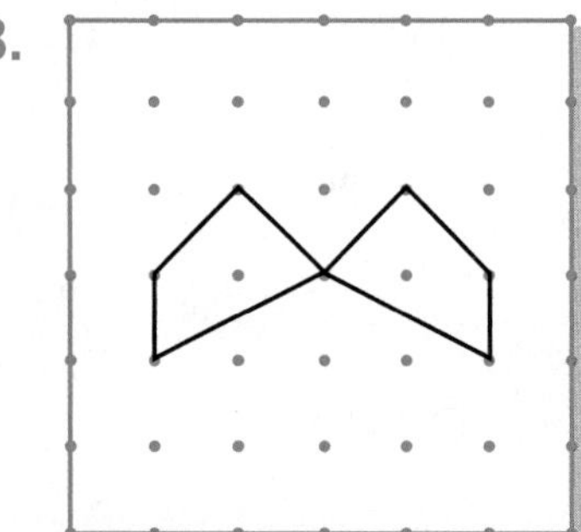

9. Which figures in Exercises 1–6 have more than one line of symmetry?

10. Choose one of the figures from your answer to Exercise 9. Draw the figure and draw all of its lines of symmetry.

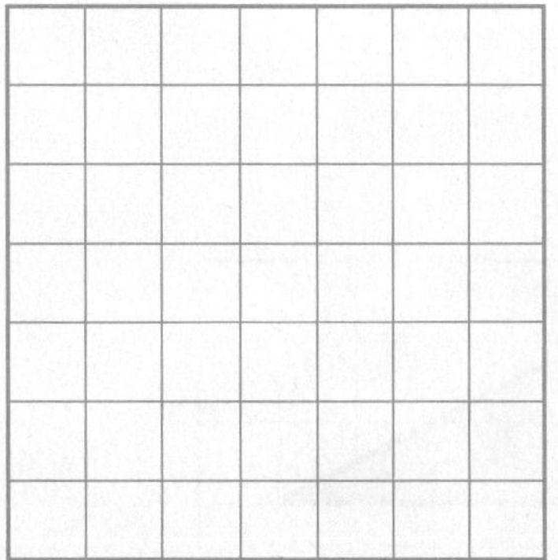

► What's the Error?

Dear Math Students:

I drew the diagonal of this rectangle as a line of symmetry.

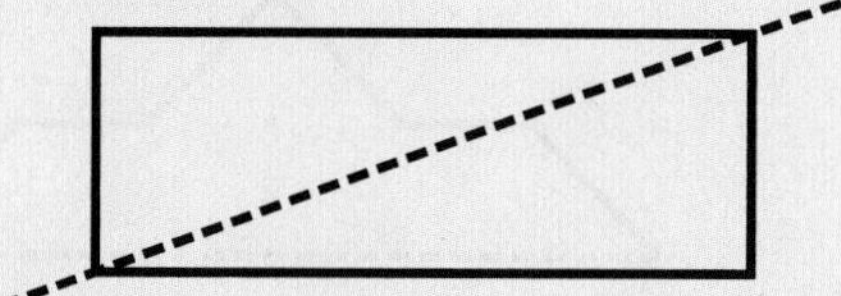

My friend told me I made a mistake. Can you help me figure out what my mistake was?

Your friend,
Puzzled Penguin

11. Write a response to Puzzled Penguin.

Use Activity Workbook page 118.

► Draw the Other Half

Draw the other half of each figure to make a whole figure or design with line symmetry.

12.

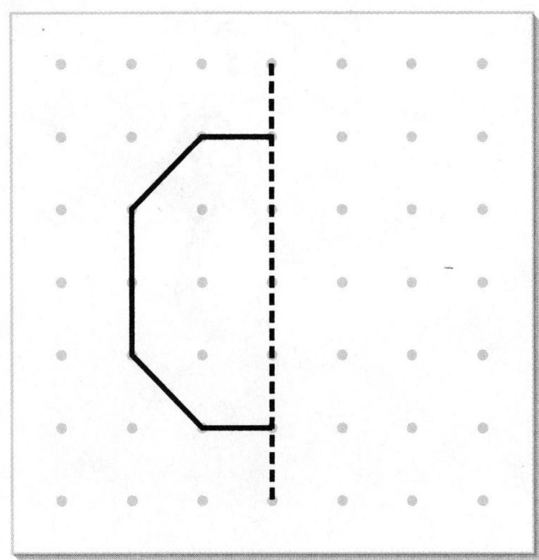

13.

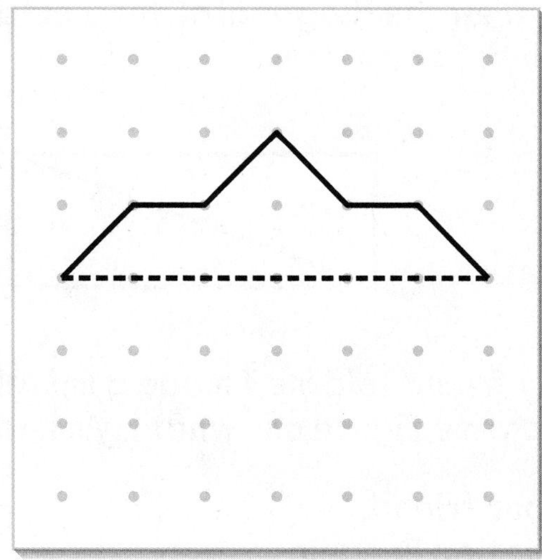

14.

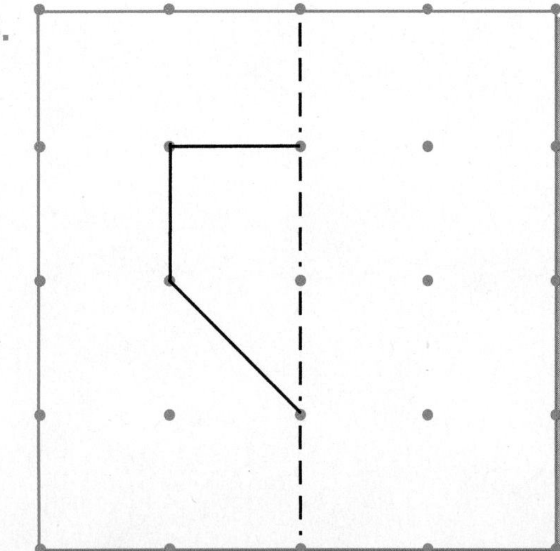

15.

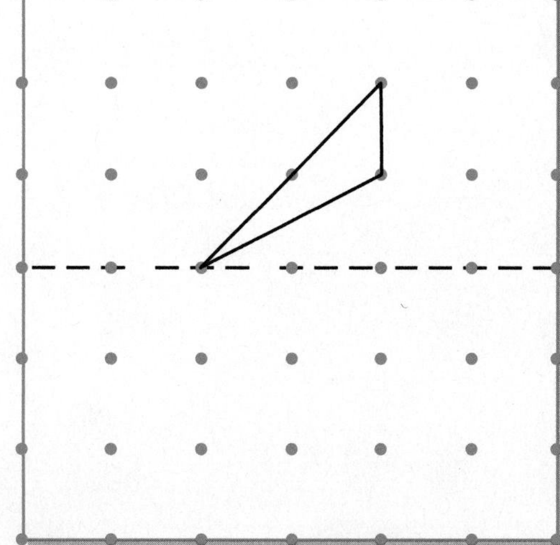

16. **Check Your Work** Copy one of your answers to Exercises 12–15 onto another piece of paper. Cut out the design and then fold it along the line of symmetry. Check that the two halves of the design match exactly.

► Math and Flags of the World

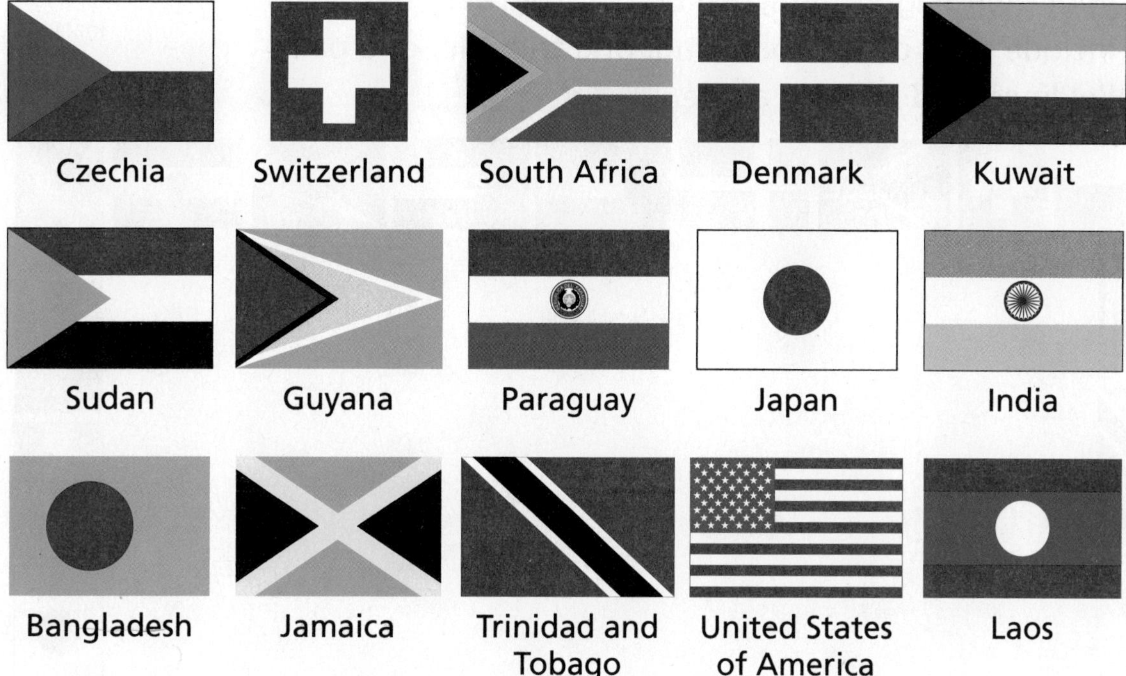

Flags are used in many different ways. Some sports teams use flags to generate team spirit, a flag might be used to start a race, or a homeowner might use a flag for decoration. States and countries also use flags as a representation of their communities. Each flag is different, both in color and design.

Use the designs on the flags to answer the questions.

1. What types of quadrilaterals are used in the Kuwait flag?

2. How many designs have no parallel lines? Name the flags.

3. How many designs have perpendicular lines? Name the flags.

4. Which designs have at least two lines of symmetry?

Show your work on your paper or in your journal.

▶ Designer Flags

Design your own flag in the space below. Your flag design should include each of the following: one triangle, one pair of parallel lines, and one 30° angle.

5. What type of triangle did you draw in your flag design? Explain how the sides of the triangle helped you classify the triangle.

6. Compare the flag design you made to the flag design that a classmate made. How are the two designs the same? How are they different? What shapes did you use that your classmate did not use?

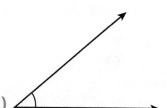

Use the Activity Workbook Unit Test on pages 119–122.

► Vocabulary

VOCABULARY
acute angle
degree
parallel
perpendicular

Choose the best term from the box.

1. A ______ is $\frac{1}{360}$th of a circle. (Lesson 8-2)
2. Two lines are ______ if they form a right angle. (Lesson 8-7)
3. An ______ has a measure less than 90°. (Lesson 8-1)

► Concepts and Skills

4. Explain how you would use a protractor to measure the angle at the right. What is the angle measure? (Lesson 8-2)

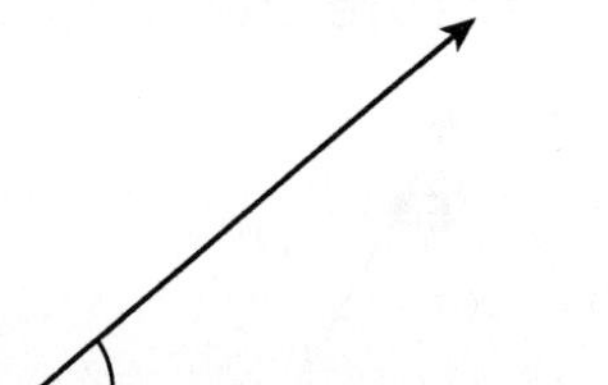

5. Look at the figures below. Circle the figures that have parallel lines. (Lesson 8-10)

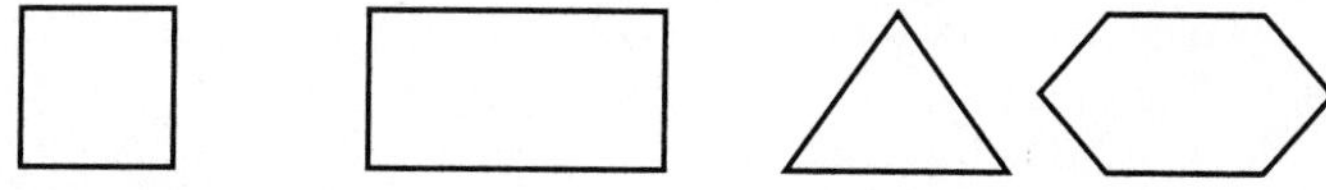

6. Look at the figures below. Circle the figures that have acute angles. (Lesson 8-10)

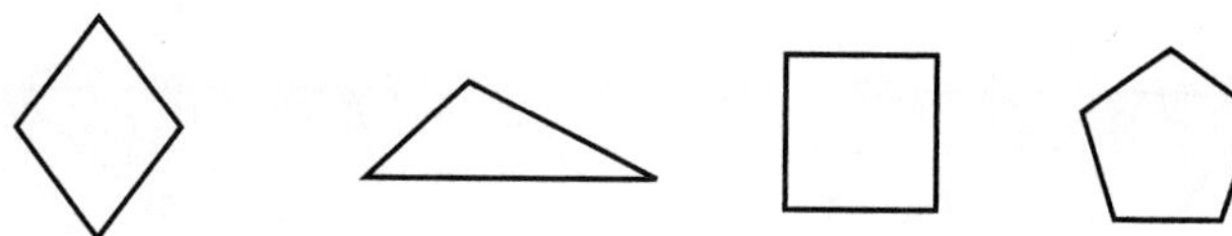

Draw each figure. (Lesson 8-1)

7. Line *AB*

8. Line segment *FG*

Tell whether each pair of lines is parallel or perpendicular. (Lesson 8-7)

9.

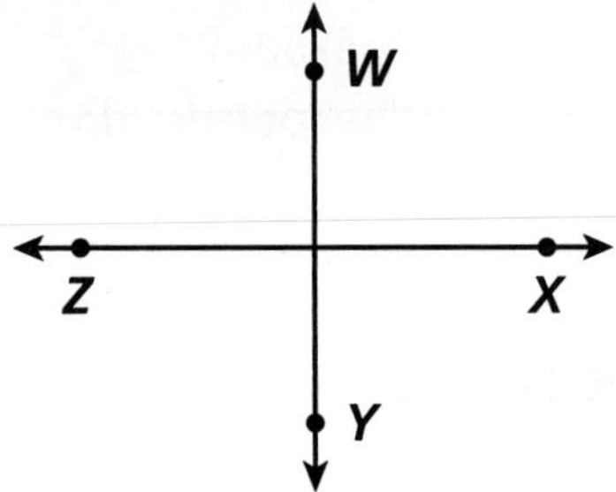

10.

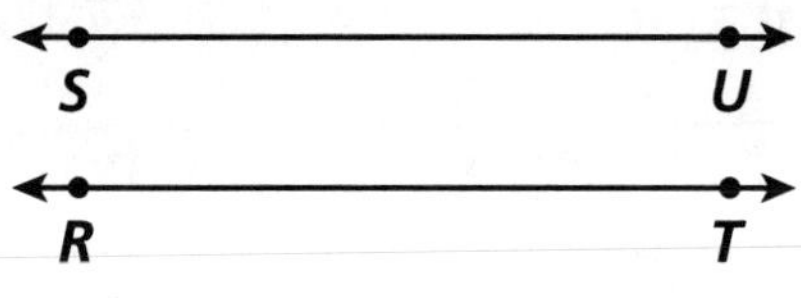

Measure the angle. Tell if it is an acute, obtuse, or right angle. (Lesson 8-2)

11.

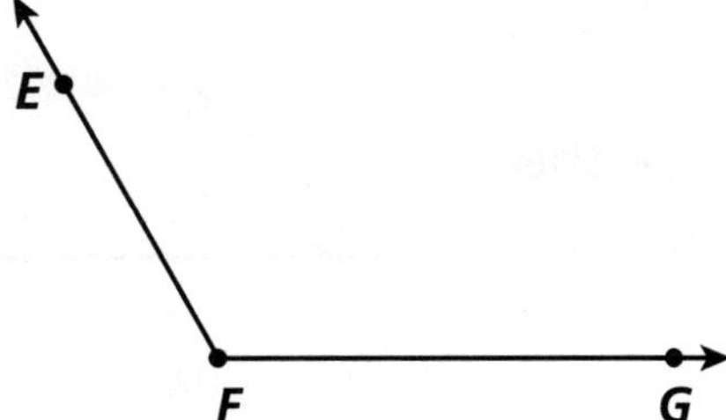

12.

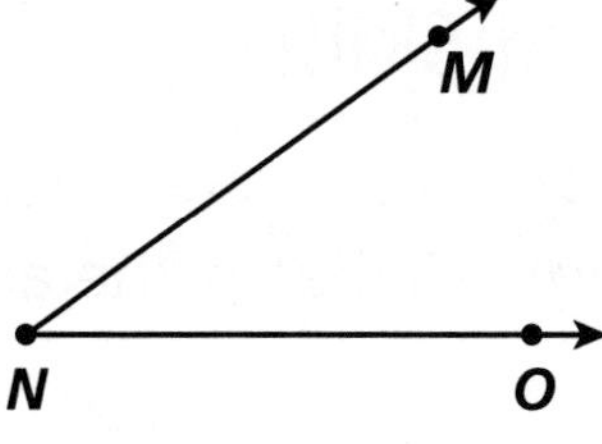

Name each triangle by its sides. (Lesson 8-4)

13.

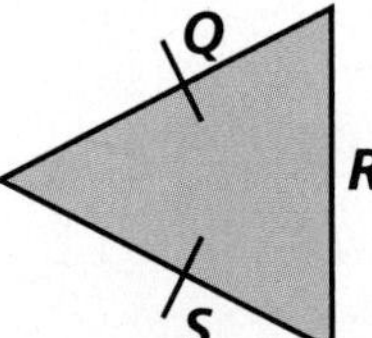

14.

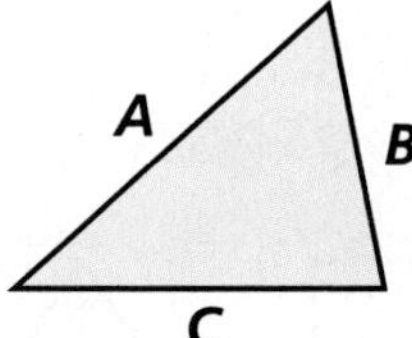

List all names for the quadrilateral. Then use letters to name the triangles you can make with the diagonals and write the type of triangles. (Lessons 8-8, 8-9)

15.

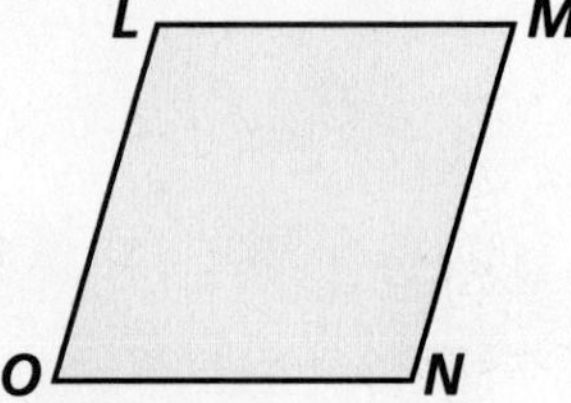

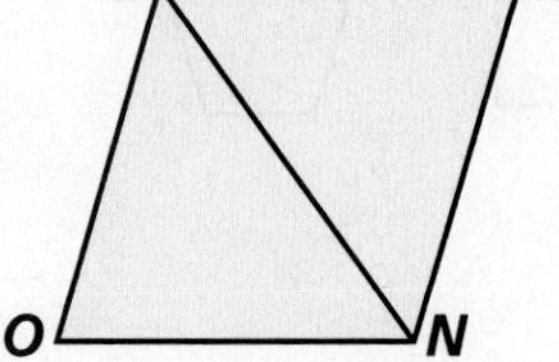

Draw all of the lines of symmetry for each figure. (Lesson 8-11)

16.

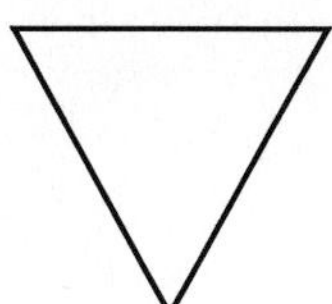

17.

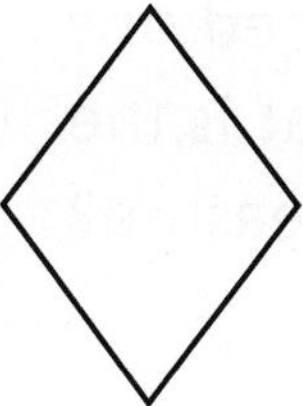

▶ Problem Solving

Use the map to solve each problem. (Lessons 8-4, 8-6, 8-7)

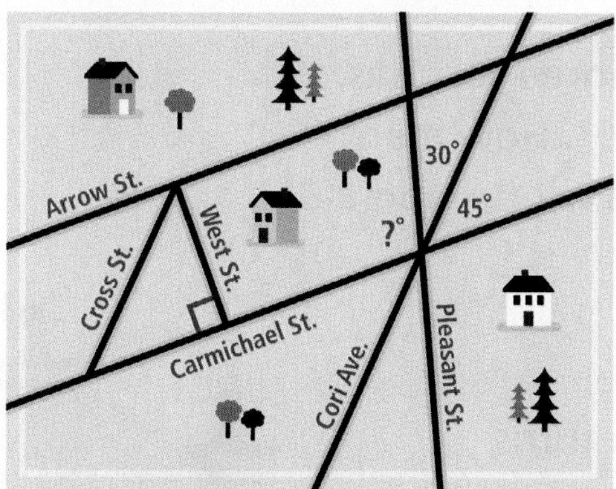

18. Suli and Ty are walking along parallel streets. Which two streets in the map appear to be parallel?

19. Cross Street, West Street, and Carmichael Street form a triangle around a park. Classify the triangle formed by these streets by its sides and its angles.

20. What is the measure of the obtuse angle formed by Pleasant Street and Carmichael Street?

21. Which two streets are perpendicular?

Solve each problem.

22. Lucy is designing a block for a quilt. She measured one of the angles. What is the unknown angle measure? (Lessons 8-5, 8-6)

23. A tile has two pairs of parallel sides and two pairs of equal sides. What shape is the tile? (Lesson 8-8)

24. A gear in a watch turns in one-degree sections. The gear has turned a total of 300°. How many one-degree turns did the gear make? (Lesson 8-3)

25. **Extended Response** A Ferris wheel turns 35° before it pauses. It then turns another 85° before stopping again. What is the total measure of the angle that the Ferris wheel turned? How many more times will it need to repeat the pattern to turn 360°? Explain your thinking. (Lessons 8-2, 8-3, 8-5, 8-6)

Reference Tables

Table of Measures	
Metric	**Customary**
Length/Area	
1,000 millimeters (mm) = 1 meter (m)	1 foot (ft) = 12 inches (in.)
100 centimeters (cm) = 1 meter	1 yard (yd) = 36 inches
10 decimeters (dm) = 1 meter	1 yard = 3 feet
1 dekameter (dam) = 10 meters	1 mile (mi) = 5,280 feet
1 hectometer (hm) = 100 meters	1 mile = 1,760 yards
1 kilometer (km) = 1,000 meters	
Liquid Volume	
1,000 milliliters (mL) = 1 liter (L)	6 teaspoons (tsp) = 1 fluid ounce (fl oz)
100 centiliters (cL) = 1 liter	2 tablespoons (tbsp) = 1 fluid ounce
10 deciliters (dL) = 1 liter	1 cup (c) = 8 fluid ounces
1 dekaliter (daL) = 10 liters	1 pint (pt) = 2 cups
1 hectoliter (hL) = 100 liters	1 quart (qt) = 2 pints
1 kiloliter (kL) = 1,000 liters	1 gallon (gal) = 4 quarts
Mass	**Weight**
1,000 milligrams (mg) = 1 gram (g)	1 pound (lb) = 16 ounces
100 centigrams (cg) = 1 gram	1 ton (T) = 2,000 pounds
10 decigrams (dg) = 1 gram	
1 dekagram (dag) = 10 grams	
1 hectogram (hg) = 100 grams	
1 kilogram (kg) = 1,000 grams	
1 metric ton = 1,000 kilograms	

Table of Units of Time

Time

1 minute (min) = 60 seconds (sec)
1 hour (hr) = 60 minutes
1 day = 24 hours
1 week (wk) = 7 days
1 month, about 30 days
1 year (yr) = 12 months (mo) or about 52 weeks

1 year = 365 days
1 leap year = 366 days
1 decade = 10 years
1 century = 100 years
1 millennium = 1,000 years

Table of Formulas

Perimeter

Polygon
$P =$ sum of the lengths of the sides

Rectangle
$P = 2(l + w)$ or $P = 2l + 2w$

Square
$P = 4s$

Area

Rectangle
$A = lw$ or $A = bh$

Square
$A = s \cdot s$

Properties of Operations

Associative Property of Addition

$(a + b) + c = a + (b + c)$ $(2 + 5) + 3 = 2 + (5 + 3)$

Commutative Property of Addition

$a + b = b + a$ $4 + 6 = 6 + 4$

Addition Identity Property of 0

$a + 0 = 0 + a = a$ $3 + 0 = 0 + 3 = 3$

Associative Property of Multiplication

$(a \cdot b) \cdot c = a \cdot (b \cdot c)$ $(3 \cdot 5) \cdot 7 = 3 \cdot (5 \cdot 7)$

Commutative Property of Multiplication

$a \cdot b = b \cdot a$ $6 \cdot 3 = 3 \cdot 6$

Multiplicative Identity Property of 1

$a \cdot 1 = 1 \cdot a = a$ $8 \cdot 1 = 1 \cdot 8 = 8$

Distributive Property of Multiplication over Addition

$a \cdot (b + c) = (a \cdot b) + (a \cdot c)$ $2 \cdot (4 + 3) = (2 \cdot 4) + (2 \cdot 3)$

Problem Types

Addition and Subtraction Problem Types

	Result Unknown	Change Unknown	Start Unknown
Add to	A glass contained $\frac{3}{4}$ cup of orange juice. Then $\frac{1}{4}$ cup of pineapple juice was added. How much juice is in the glass now? *Situation and solution equation:* [1] $\frac{3}{4} + \frac{1}{4} = c$	A glass contained $\frac{3}{4}$ cup of orange juice. Then some pineapple juice was added. Now the glass contains 1 cup of juice. How much pineapple juice was added? *Situation equation:* $\frac{3}{4} + c = 1$ *Solution equation:* $c = 1 - \frac{3}{4}$	A glass contained some orange juice. Then $\frac{1}{4}$ cup of pineapple juice was added. Now the glass contains 1 cup of juice. How much orange juice was in the glass to start? *Situation equation* $c + \frac{1}{4} = 1$ *Solution equation:* $c = 1 - \frac{1}{4}$
Take from	Micah had a ribbon $\frac{5}{6}$ yard long. He cut off a piece $\frac{1}{6}$ yard long. What is the length of the ribbon that is left? *Situation and solution equation:* $\frac{5}{6} - \frac{1}{6} = r$	Micah had a ribbon $\frac{5}{6}$ yard long. He cut off a piece. Now the ribbon is $\frac{4}{6}$ yard long. What is the length of the ribbon he cut off? *Situation equation:* $\frac{5}{6} - r = \frac{4}{6}$ *Solution equation:* $r = \frac{5}{6} - \frac{4}{6}$	Micah had a ribbon. He cut off a piece $\frac{1}{6}$ yard long. Now the ribbon is $\frac{4}{6}$ yard long. What was the length of the ribbon he started with? *Situation equation:* $r - \frac{1}{6} = \frac{4}{6}$ *Solution equation:* $r = \frac{4}{6} + \frac{1}{6}$

[1]A situation equation represents the structure (action) in the problem situation. A solution equation shows the operation used to find the answer.

	Total Unknown	Addend Unknown	Other Addend Unknown
Put Together/ Take Apart	A baker combines $1\frac{2}{3}$ cups of white flour and $\frac{2}{3}$ cup of wheat flour. How much flour is this altogether? *Math drawing:*[1] 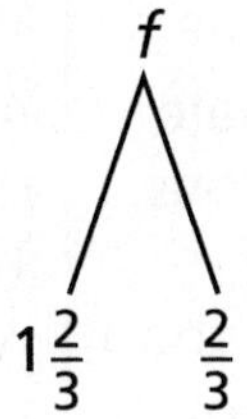 f breaks apart into $1\frac{2}{3}$ and $\frac{2}{3}$ *Situation and solution equation:* $1\frac{2}{3} + \frac{2}{3} = f$	Of the $2\frac{1}{3}$ cups of flour a baker uses, $1\frac{2}{3}$ cups are white flour. The rest is wheat flour. How much wheat flour does the baker use? *Math drawing:*  $2\frac{1}{3}$ breaks apart into $1\frac{2}{3}$ and f *Situation equation:* $2\frac{1}{3} = 1\frac{2}{3} + f$ *Solution equation:* $f = 2\frac{1}{3} - 1\frac{2}{3}$	A baker uses $2\frac{1}{3}$ cups of flour. Some is white flour and $\frac{2}{3}$ cup is wheat flour. How much white flour does the baker use? *Math drawing:* 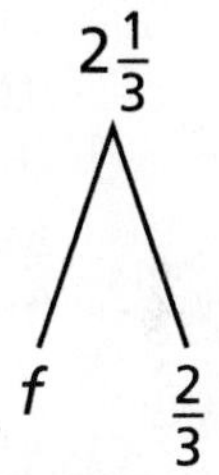 $2\frac{1}{3}$ breaks apart into f and $\frac{2}{3}$ *Situation equation* $2\frac{1}{3} = f + \frac{2}{3}$ *Solution equation:* $f = 2\frac{1}{3} - \frac{2}{3}$

[1]These math drawings are called math mountains in Grades 1–3 and break apart drawings in Grades 4 and 5.

Problem Types continued

Addition and Subtraction Problem Types (continued)

	Difference Unknown	Greater Unknown	Smaller Unknown
Additive Comparison[1]	At a zoo, the female rhino weighs $1\frac{3}{5}$ tons. The male rhino weighs $2\frac{2}{5}$ tons. How much more does the male rhino weigh than the female rhino? At a zoo, the female rhino weighs $1\frac{3}{5}$ tons. The male rhino weighs $2\frac{2}{5}$ tons. How much less does the female rhino weigh than the male rhino? *Math drawing:* $2\frac{2}{5}$ $1\frac{3}{5}$ d *Situation equation:* $1\frac{3}{5} + d = 2\frac{2}{5}$ or $d = 2\frac{2}{5} - 1\frac{3}{5}$ *Solution equation:* $d = 2\frac{2}{5} - 1\frac{3}{5}$	**Leading Language** At a zoo, the female rhino weighs $1\frac{3}{5}$ tons. The male rhino weighs $\frac{4}{5}$ ton more than the female rhino. How much does the male rhino weigh? **Misleading Language** At a zoo, the female rhino weighs $1\frac{3}{5}$ tons. The female rhino weighs $\frac{4}{5}$ ton less than the male rhino. How much does the male rhino weigh? *Math drawing:* 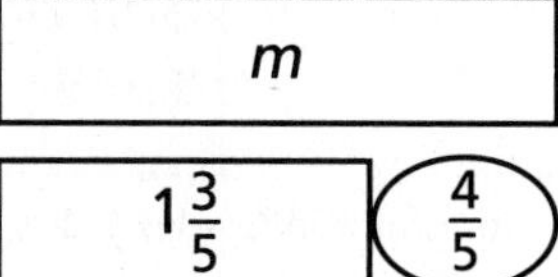 *Situation and solution equation:* $1\frac{3}{5} + \frac{4}{5} = m$	**Leading Language** At a zoo, the male rhino weighs $2\frac{2}{5}$ tons. The female rhino weighs $\frac{4}{5}$ ton less than the male rhino. How much does the female rhino weigh? **Misleading Language** At a zoo, the male rhino weighs $2\frac{2}{5}$ tons. The male rhino weighs $\frac{4}{5}$ ton more than the female rhino. How much does the female rhino weigh? *Math drawing:* $2\frac{2}{5}$ f $\frac{4}{5}$ *Situation equation* $f + \frac{4}{5} = 2\frac{2}{5}$ or $f = 2\frac{2}{5} - \frac{4}{5}$ *Solution equation:* $f = 2\frac{2}{5} - \frac{4}{5}$

[1]A comparison sentence can always be said in two ways. One way uses *more*, and the other uses *fewer* or *less*. Misleading language suggests the wrong operation. For example, it says *the female rhino weighs $\frac{4}{5}$ ton less than the male*, but you have to add $\frac{4}{5}$ ton to the female's weight to get the male's weight.

Multiplication and Division Problem Types

	Unknown Product	Group Size Unknown	Number of Groups Unknown
Equal Groups	A teacher bought 10 boxes of pencils. There are 20 pencils in each box. How many pencils did the teacher buy? *Situation and solution equation:* $p = 10 \cdot 20$	A teacher bought 10 boxes of pencils. She bought 200 pencils in all. How many pencils are in each box? *Situation equation:* $10 \cdot n = 200$ *Solution equation:* $n = 200 \div 10$	A teacher bought boxes of 20 pencils. She bought 200 pencils in all. How many boxes of pencils did she buy? *Situation equation* $b \cdot 20 = 200$ *Solution equation:* $b = 200 \div 20$

	Unknown Product	Unknown Factor	Unknown Factor
Arrays[1]	An auditorium has 60 rows with 30 seats in each row. How many seats are in the auditorium? *Math drawing:* 30 60 [s] *Situation and solution equation:* $s = 60 \cdot 30$	An auditorium has 60 rows with the same number of seats in each row. There are 1,800 seats in all. How many seats are in each row? *Math drawing:* *n* 60 [1,800] *Situation equation:* $60 \cdot n = 1{,}800$ *Solution equation:* $n = 1{,}800 \div 60$	The 1,800 seats in an auditorium are arranged in rows of 30. How many rows of seats are there? *Math drawing:* 30 *r* [1,800] *Situation equation* $r \cdot 30 = 1{,}800$ *Solution equation:* $r = 1{,}800 \div 30$

[1]We use rectangle models for both array and area problems in Grades 4 and 5 because the numbers in the problems are too large to represent with arrays.

Problem Types (continued)

Multiplication and Division Problem Types (continued)

	Unknown Product	Unknown Factor	Unknown Factor
Area	Sophie's backyard is 80 feet long and 40 feet wide. What is the area of Sophie's backyard? *Math drawing:* 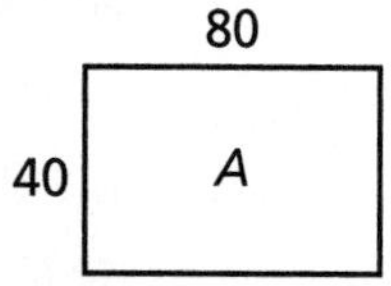 80 40 A *Situation and solution equation:* $A = 80 \cdot 40$	Sophie's backyard has an area of 3,200 square feet. The length of the yard is 80 feet. What is the width of the yard? *Math drawing:* 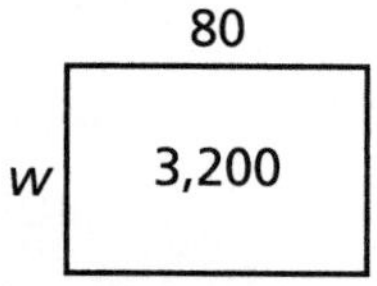 80 w 3,200 *Situation equation:* $80 \cdot w = 3{,}200$ *Solution equation:* $w = 3{,}200 \div 80$	Sophie's backyard has an area of 3,200 square feet. The width of the yard is 40 feet. What is the length of the yard? *Math drawing:* 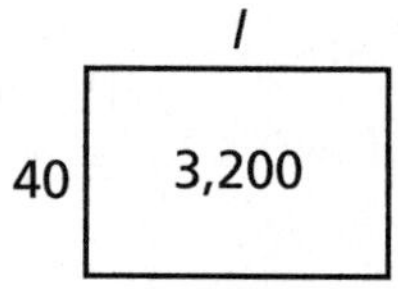 l 40 3,200 *Situation equation* $l \cdot 40 = 3{,}200$ *Solution equation:* $l = 3{,}200 \div 40$
Multiplicative Comparison	**Whole Number Multiplier** Sam has 4 times as many marbles as Brady has. Brady has 70 marbles. How many marbles does Sam have? *Math drawing:* 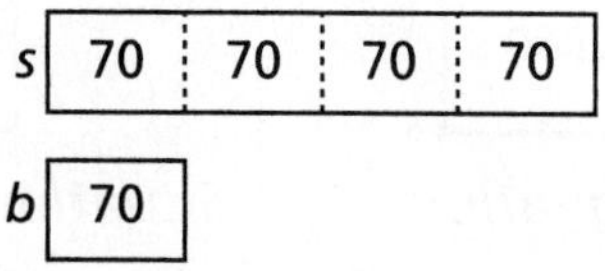 s 70 70 70 70 b 70 *Situation and solution equation:* $s = 4 \cdot 70$	**Whole Number Multiplier** Sam has 4 times as many marbles as Brady has. Sam has 280 marbles. How many marbles does Brady have? *Math drawing:* 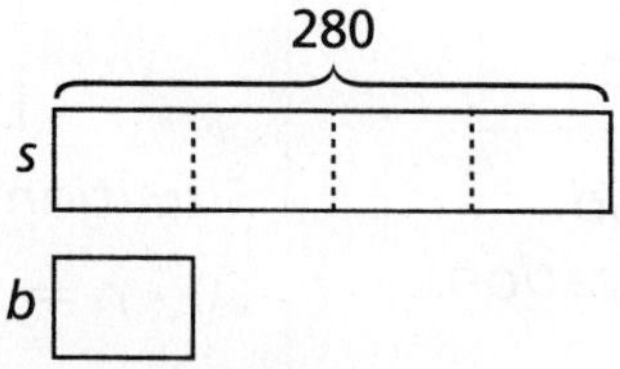 280 s b *Situation equation:* $4 \cdot b = 280$ *Solution equation:* $b = 280 \div 4$	**Whole Number Multiplier** Sam has 280 marbles. Brady has 70 marbles. The number of marbles Sam has is how many times the number Brady has? *Math drawing:* 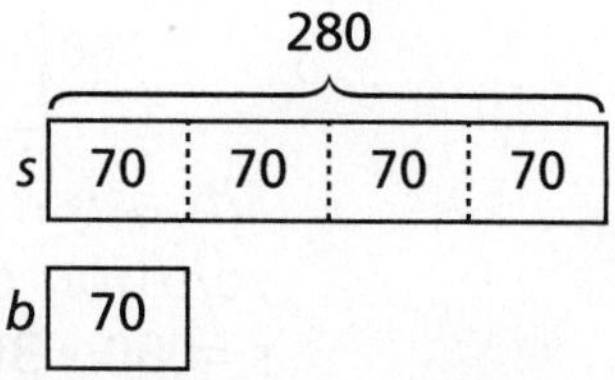 280 s 70 70 70 70 b 70 *Situation equation* $m \cdot 70 = 280$ *Solution equation:* $m = 280 \div 70$

Vocabulary Activities

MathWord Power

► Word Review PAIRS

Work with a partner. Choose a word from a current unit or a review word from a previous unit. Use the word to complete one of the activities listed on the right. Then ask your partner if they have any edits to your work or questions about what you described. Repeat, having your partner choose a word.

Activities

- Give the meaning in words or gestures.
- Use the word in the sentence.
- Give another word that is related to the word in some way and explain the relationship.

► Crossword Puzzle PAIRS OR INDIVIDUALS

Create a crossword puzzle similar to the example below. Use vocabulary words from the unit. You can add other related words, too. Challenge your partner to solve the puzzle.

	1 d						
	i			2 s	u	m	
	v			u			
	i			b			3 u
4 a	d	d	i	t	i	o	n
	e			r			i
	n			a			t
5 a	d	d		c			
				t			
				i			
		6 g	r	o	u	p	
				n			

Across

2. The answer to an addition problem
4. _____ and subtraction are inverse operations.
5. To put amounts together
6. When you trade 10 ones for 1 ten, you _____.

Down

1. The number to be divided in a division problem
2. The operation that you can use to find out how much more one number is than another.
3. A fraction with a numerator of 1 is a _____ fraction.

▶ Word Wall PAIRS OR SMALL GROUPS

With your teacher's permission, start a word wall in your classroom. As you work through each lesson, put the math vocabulary words on index cards and place them on the word wall. You can work with a partner or a small group choosing a word and giving the definition.

▶ Word Web INDIVIDUALS

Make a word web for a word or words you do not understand in a unit. Fill in the web with words or phrases that are related to the vocabulary word.

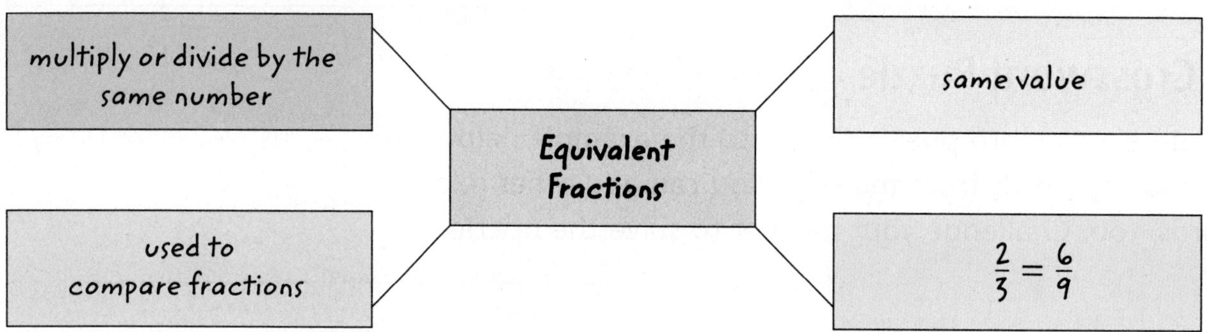

▶ Alphabet Challenge PAIRS OR INDIVIDUALS

Take an alphabet challenge. Choose 3 letters from the alphabet. Think of three vocabulary words for each letter. Then write the definition or draw an example for each word.

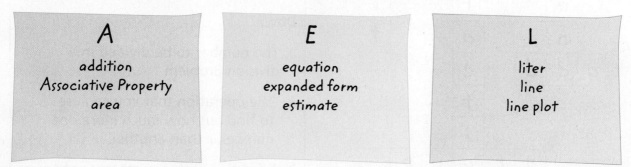

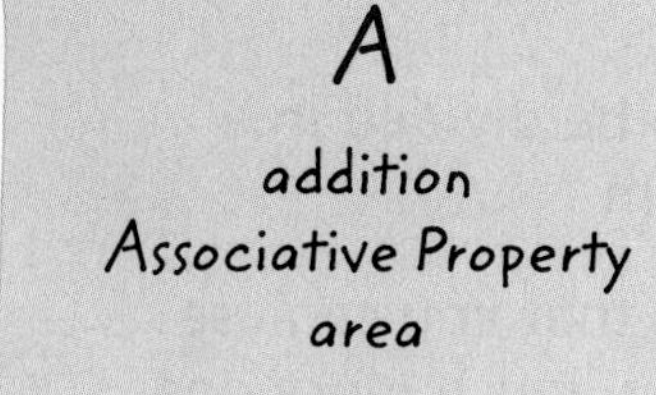

E
equation
expanded form
estimate

L
liter
line
line plot

▶ Concentration PAIRS

Write the vocabulary words and related words from a unit on index cards. Write the definitions on a different set of index cards. Mix up both sets of cards. Then place the cards facedown on a table in an array, for example, 3 by 3 or 3 by 4. Take turns turning over two cards. If one card is a word and one card is a definition that matches the word, take the pair. Continue until each word has been matched with its definition.

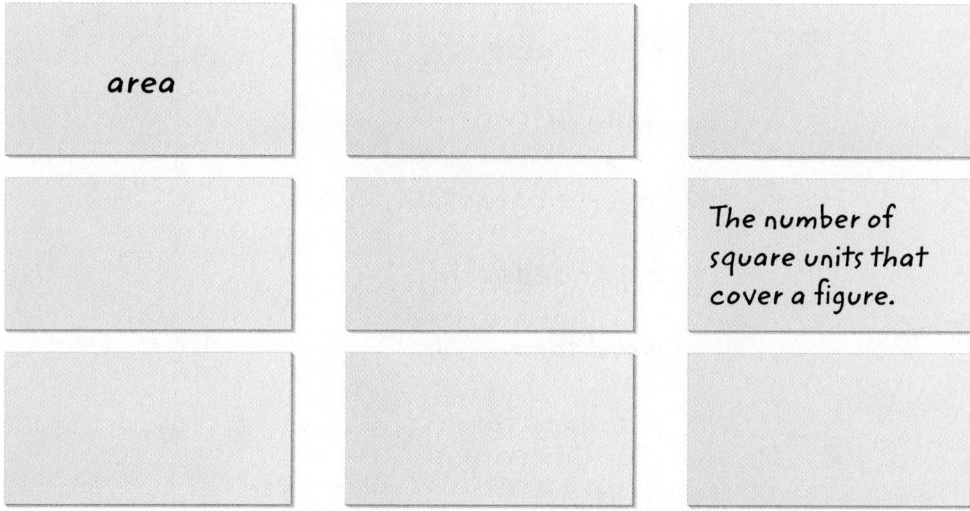
area
The number of square units that cover a figure.

▶ Math Journal INDIVIDUALS

As you learn new words, write them in your Math Journal. Write the definition of the word and include a sketch or an example. As you learn new information about the word, add notes to your definition.

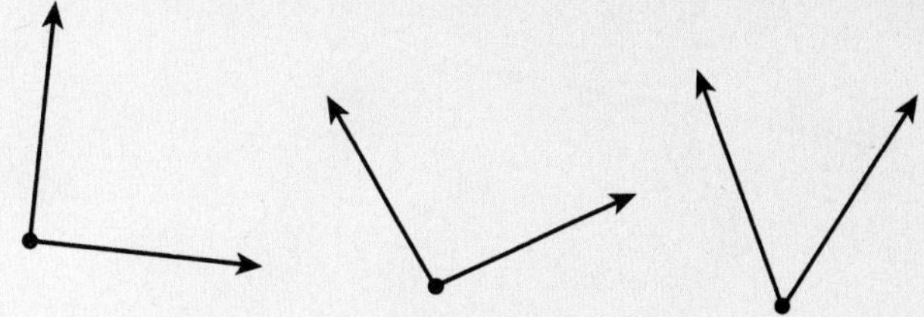
Angle: A figure formed by two rays with the same endpoint.
Degree: A unit for measuring angles.

▶ What's the Word? PAIRS

Work together to make a poster or bulletin board display of the words in a unit. Write definitions on a set of index cards. Mix up the cards. Work with a partner, choosing a definition from the index cards. Have your partner point to the word on the poster and name the matching math vocabulary word. Switch roles and try the activity again.

array

place value

addend

inverse operations

expanded form

word form

standard form

digit

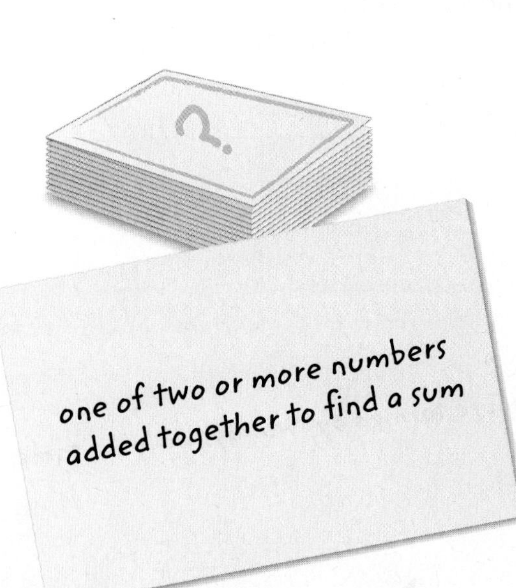
?
one of two or more numbers
added together to find a sum

Glossary

A

acute angle An angle smaller than a right angle.

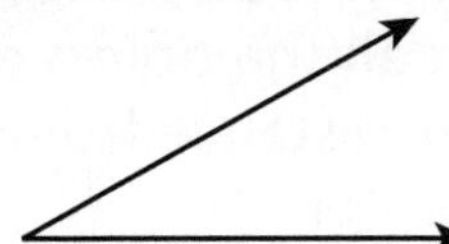

acute triangle A triangle with three acute angles.

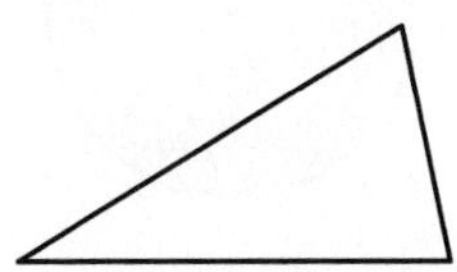

addend One of two or more numbers added together to find a sum.

Example: 7 + 8 = 15

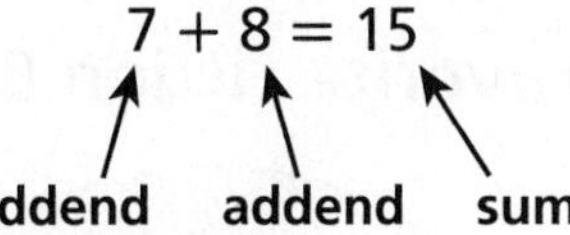

adjacent (sides) Two sides that meet at a point.

Example: Sides *a* and *b* are adjacent.

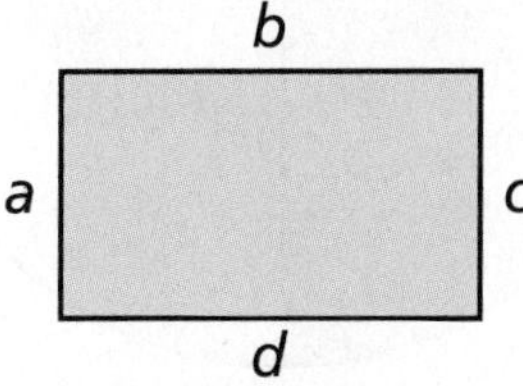

Algebraic Notation Method A strategy based on the Distributive Property in which a factor is decomposed to create simpler algebraic expressions, and the Distributive Property is applied.

Example: $9 \cdot 28 = 9 \cdot (20 + 8)$
$= (9 \cdot 20) + (9 \cdot 8)$
$= 180 + 72$
$= 252$

analog clock A clock with a face and hands.

angle A figure formed by two rays with the same endpoint.

area The number of square units that cover a figure.

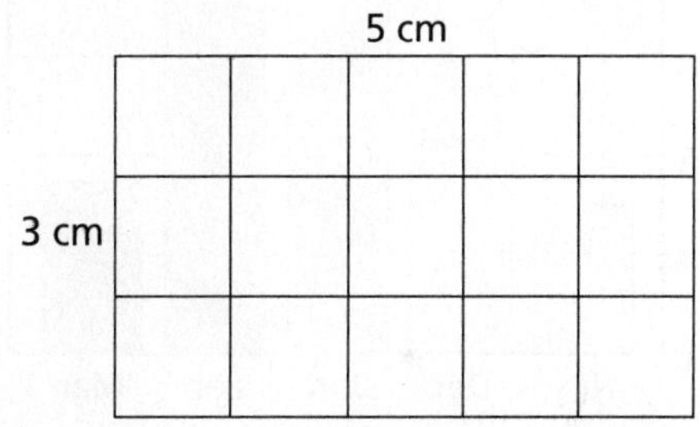

array An arrangement of objects, symbols, or numbers in rows and columns.

Associative Property of Addition Grouping the addends in different ways does not change the sum.

Example: 3 + (5 + 7) = 15
(3 + 5) + 7 = 15

Associative Property of Multiplication Grouping the factors in different ways does not change the product.

Example: $3 \times (5 \times 7) = 105$
$(3 \times 5) \times 7 = 105$

B

bar graph A graph that uses bars to show data. The bars may be vertical or horizontal.

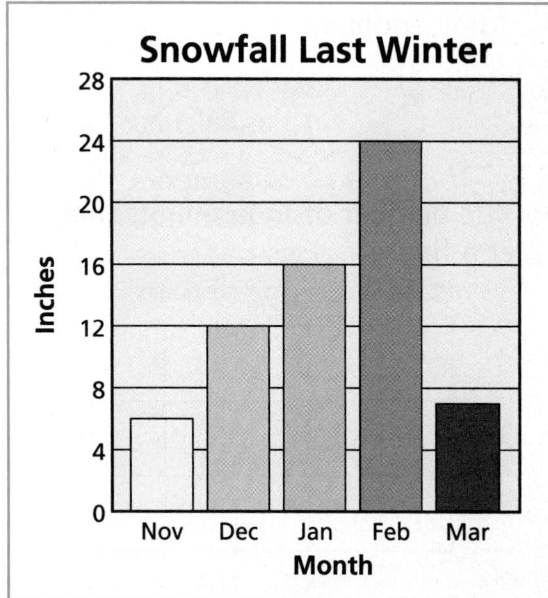

break-apart drawing A diagram that shows two addends and the sum.

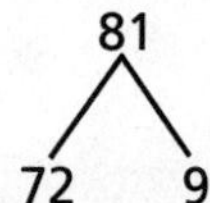

C

center The point that is the same distance from every point on the circle.

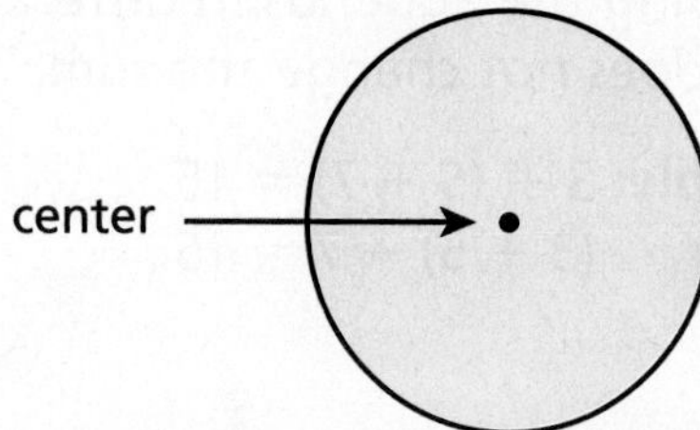

centimeter A unit of measure in the metric system that equals one hundredth of a meter. 100 cm = 1 m

circle A plane figure that forms a closed path so that all the points on the path are the same distance from a point called the center.

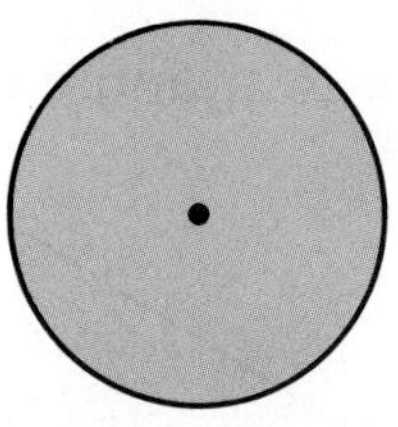

circle graph A graph that uses parts of a circle to show data.

Example:

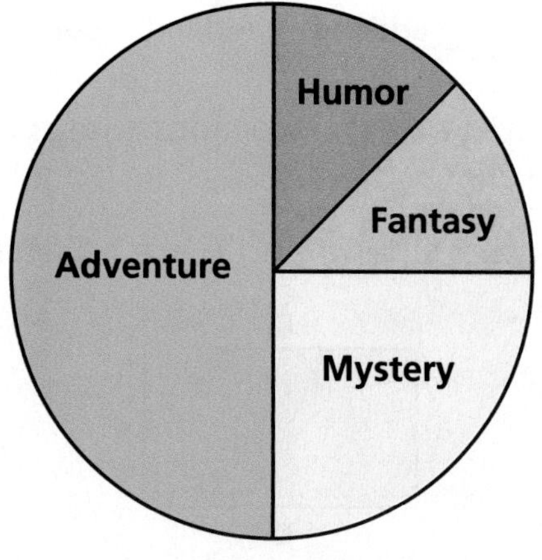

column A part of a table or array that contains items arranged vertically.

common denominator A common multiple of two or more denominators.

Example: A common denominator of $\frac{1}{2}$ and $\frac{1}{3}$ is 6 because 6 is a multiple of 2 and 3.

Commutative Property of Addition Changing the order of addends does not change the sum.

Example: $3 + 8 = 11$
$8 + 3 = 11$

Commutative Property of Multiplication Changing the order of factors does not change the product.

Example: $3 \times 8 = 24$
$8 \times 3 = 24$

compare Describe quantities as greater than, less than, or equal to each other.

comparison bars Bars that represent the larger amount and smaller amount in a comparison situation.

For addition and subtraction:

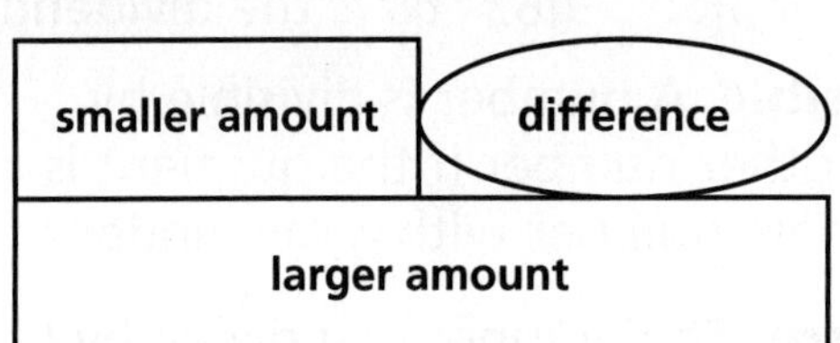

For multiplication and division:

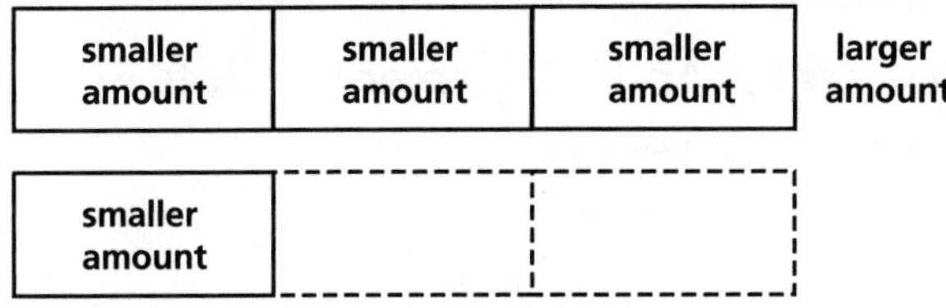

comparison situation A situation in which two amounts are compared by addition or by multiplication. An *addition comparison situation* compares by asking or telling how much more (how much less) one amount is than another. A *multiplication comparison situation* compares by asking or telling how many times as many one amount is as another. The multiplication comparison may also be made using fraction language. For example, you can say, "Sally has one fourth as much as Tom has," instead of saying "Tom has 4 times as much as Sally has."

composite number A number greater than 1 that has more than one factor pair. Examples of composite numbers are 10 and 18. The factor pairs of 10 are 1 and 10, 2 and 5. The factor pairs of 18 are 1 and 18, 2 and 9, 3 and 6.

cup A unit of liquid volume in the customary system that equals 8 fluid ounces.

D

data A collection of information.

decimal number A representation of a number using the numerals 0 to 9, in which each digit has a value 10 times the digit to its right. A dot or **decimal point** separates the whole-number part of the number on the left from the fractional part on the right.

Examples: 1.23 and 0.3

decimal point A symbol used to separate dollars and cents in money amounts or to separate ones and tenths in decimal numbers.

Examples:

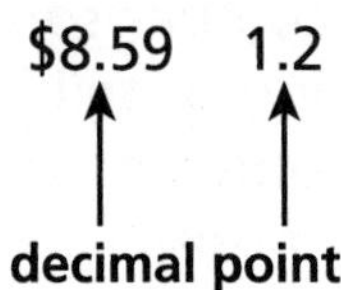

decimeter A unit of measure in the metric system that equals one tenth of a meter. 10 dm = 1 m

degree (°) A unit for measuring angles.

denominator The number below the bar in a fraction. It shows the total number of equal parts in the fraction.

Example:

$\frac{3}{4}$ ← denominator

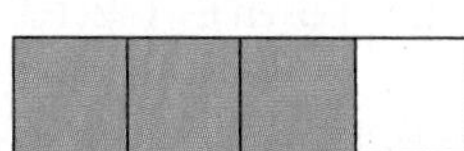

diagonal A line segment that connects vertices of a polygon, but is not a side of the polygon.

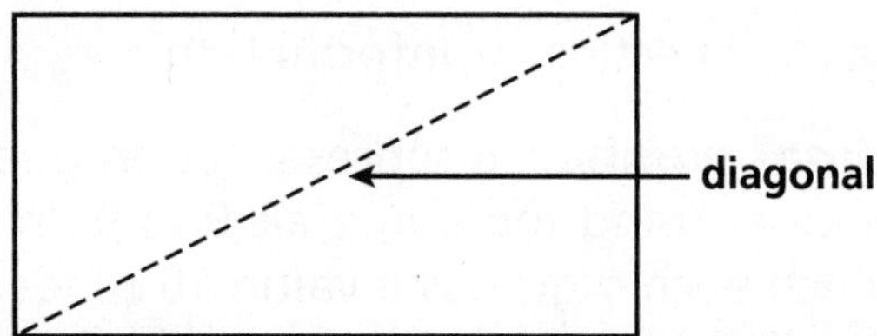

difference The result of a subtraction.

Example: 54 − 37 = 17 ← difference

digit Any of the symbols 0, 1, 2, 3, 4, 5, 6, 7, 8, or 9.

digital clock A clock that shows us the hour and minutes with numbers.

Digit-by-Digit A method used to solve a division problem.

Put in only one digit at a time.

```
    5          54          546
7)3,822    7)3,822     7)3,822
 −35        −35         −35
  32         32          32
            −28         −28
             42          42
                        −42
```

Distributive Property You can multiply a sum by a number, or multiply each addend by the number and add the products; the result is the same.

Example:

$3 \times (2 + 4) = (3 \times 2) + (3 \times 4)$

$3 \times 6 = 6 + 12$

$18 = 18$

dividend The number that is divided in division.

Example: $9\overline{)63}$ with quotient 7 — 63 is the dividend.

divisible A number is divisible by another number if the quotient is a whole number with a remainder of 0.

divisor The number you divide by in division.

Example: $9\overline{)63}$ with quotient 7 — 9 is the divisor.

dot array An arrangement of dots in rows and columns.

E

elapsed time The time that passes between the beginning and the end of an activity.

endpoint The point at either end of a line segment or the beginning point of a ray.

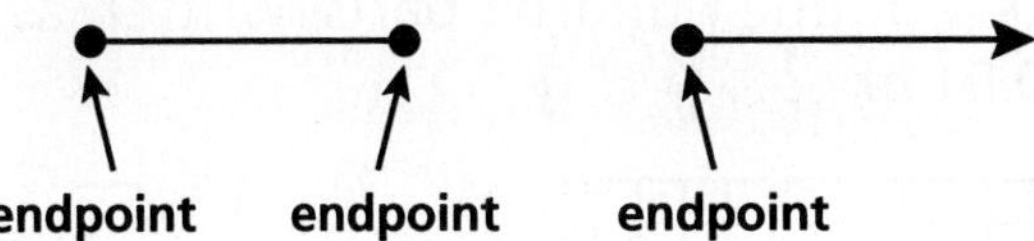

equation A statement that two expressions are equal. It has an equal sign.

Examples: $32 + 35 = 67$
$67 = 32 + 34 + 1$
$(7 \times 8) + 1 = 57$

equilateral Having all sides of equal length.

Example: An equilateral triangle

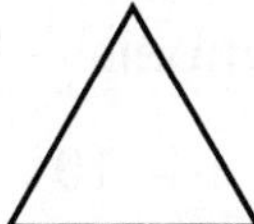

equivalent fractions Two or more fractions that represent the same number.

Example: $\frac{2}{4}$ and $\frac{4}{8}$ are equivalent because they both represent one half.

estimate A number close to an exact amount or to find about how many or how much.

evaluate Substitute a value for a letter (or symbol) and then simplify the expression.

expanded form A way of writing a number that shows the value of each of its digits.

Example: Expanded form of 835:
800 + 30 + 5
8 hundreds + 3 tens + 5 ones

Expanded Notation A method used to solve multiplication and division problems.

Examples:

43 × 67

$$\begin{array}{rcl} 67 & = & 60 + 7 \\ \times 43 & = & 40 + 3 \\ \hline 40 \times 60 & = & 2400 \\ 40 \times 7 & = & 280 \\ 3 \times 60 & = & 180 \\ 3 \times 7 & = & +\ 21 \\ \hline & & 2{,}881 \end{array}$$

3,822 ÷ 7

$$\begin{array}{r} 6 \\ 40 \\ 500 \\ 7\overline{)\,3{,}822} \\ -\,3\,500 \\ \hline 322 \\ -\,280 \\ \hline 42 \\ -\,42 \\ \hline 0 \end{array} \quad \text{6 + 40 + 500 = } 546$$

expression One or more numbers, variables, or numbers and variables with one or more operations.

Examples: 4
$6x$
$6x - 5$
$7 + 4$

F

factor One of two or more numbers multiplied to find a product.

Example: $4 \times 5 = 20$
(4: factor; 5: factor; 20: product)

factor pair A factor pair for a number is a pair of whole numbers whose product is that number.

Example: $5 \times 7 = 35$

factor pair — product

fluid ounce A unit of liquid volume in the customary system.
8 fluid ounces = 1 cup

foot A U.S. customary unit of length equal to 12 inches.

formula An equation with letters or symbols that describes a rule.

The formula for the area of a rectangle is:

$A = l \times w$

where A is the area, l is the length, and w is the width.

fraction A number that is the sum of unit fractions, each an equal part of a set or part of a whole.

Examples: $\frac{3}{4} = \frac{1}{4} + \frac{1}{4} + \frac{1}{4}$

$\frac{5}{4} = \frac{1}{4} + \frac{1}{4} + \frac{1}{4} + \frac{1}{4} + \frac{1}{4}$

G

gallon A unit of liquid volume in the customary system that equals 4 quarts.

gram The basic unit of mass in the metric system.

greater than (>) A symbol used to compare two numbers. The greater number is given first below.

Example: $33 > 17$
33 is greater than 17.

group To combine numbers to form new tens, hundreds, thousands, and so on.

H

hundredth A unit fraction representing one of one hundred parts, written as 0.01 or $\frac{1}{100}$.

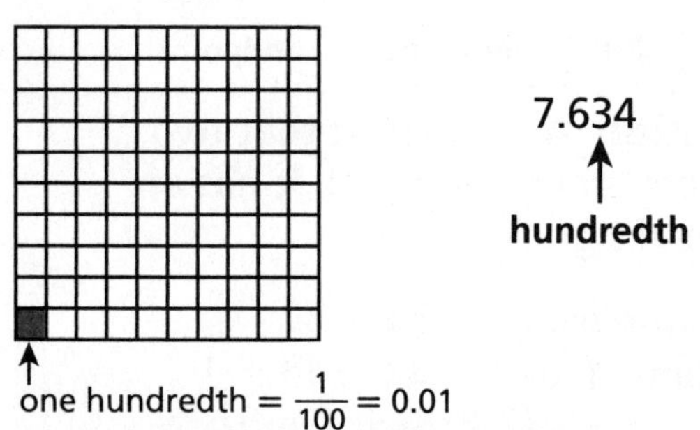

I

Identity Property of Multiplication The product of 1 and any number equals that number.

Example: $10 \times 1 = 10$

inch A U.S. customary unit of length.

Example: 1 inch

inequality A statement that two expressions are not equal.

Examples: $2 < 5$
$4 + 5 > 12 - 8$

inverse operations Opposite or reverse operations that undo each other. Addition and subtraction are inverse operations. Multiplication and division are inverse operations.

Examples: $4 + 6 = 10$ so, $10 - 6 = 4$ and $10 - 4 = 6$.
$3 \times 9 = 27$ so, $27 \div 9 = 3$ and $27 \div 3 = 9$.

isosceles triangle A triangle with at least two sides of equal length.

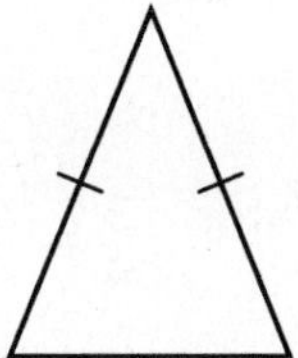

K

kilogram A unit of mass in the metric system that equals one thousand grams. 1 kg = 1,000 g

kiloliter A unit of liquid volume in the metric system that equals one thousand liters. 1 kL = 1,000 L

kilometer A unit of length in the metric system that equals 1,000 meters. 1 km = 1,000 m

L

least common denominator The least common multiple of two or more denominators.

Example: The least common denominator of $\frac{1}{2}$ and $\frac{1}{3}$ is 6 because 6 is the smallest multiple of 2 and 3.

length The measure of a line segment or one side of a figure.

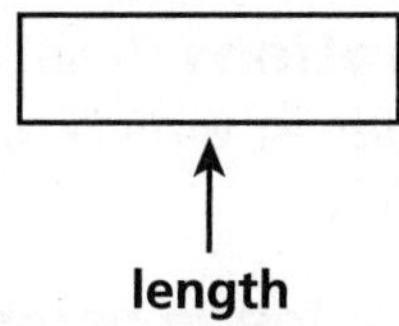

less than (<) A symbol used to compare two numbers. The smaller number is given first below.

Example: 54 < 78
54 is less than 78.

line A straight path that goes on forever in opposite directions.

Example: line *AB*

line of symmetry A line on which a figure can be folded so that the two halves match exactly.

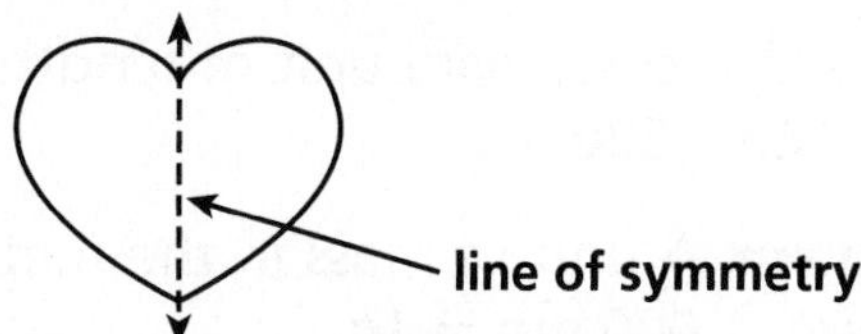

line plot A diagram that shows the frequency of data on a number line. Also called a dot plot.

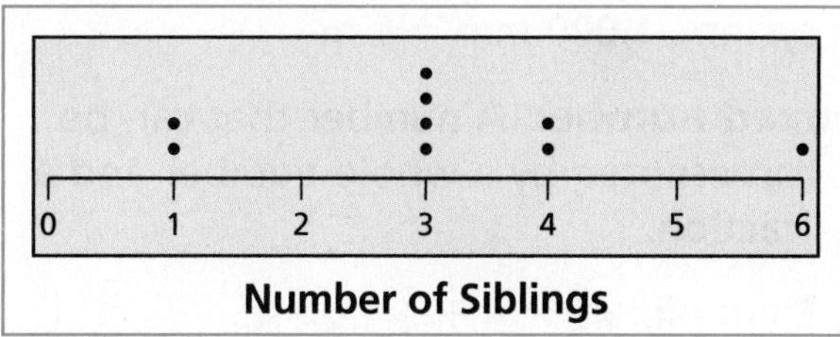

line segment Part of a line that has two endpoints.

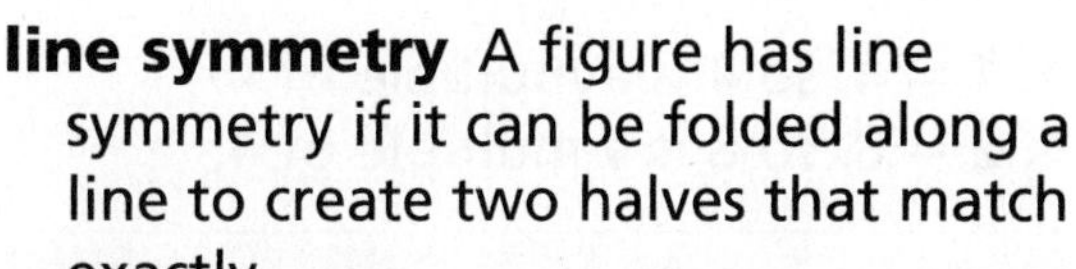

line symmetry A figure has line symmetry if it can be folded along a line to create two halves that match exactly.

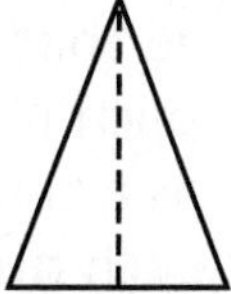

liquid volume A measure of the space a liquid occupies.

liter The basic unit of liquid volume in the metric system.
1 liter = 1,000 milliliters

M

mass The measure of the amount of matter in an object.

meter The basic unit of length in the metric system.

metric system A base ten system of measurement.

mile A U.S. customary unit of length equal to 5,280 feet.

milligram A unit of mass in the metric system. 1,000 mg = 1g

milliliter A unit of liquid volume in the metric system. 1,000 mL = 1 L

millimeter A unit of length in the metric system. 1,000 mm = 1 m

mixed number A number that can be represented by a whole number and a fraction.

Example: $4\frac{1}{2} = 4 + \frac{1}{2}$

multiple A number that is the product of a given number and any whole number.

Examples:

4 × 1 = 4, so 4 is a multiple of 4.
4 × 2 = 8, so 8 is a multiple of 4.

N

number line A line that extends, without end, in each direction and shows numbers as a series of points. The location of each number is shown by its distance from 0.

numerator The number above the bar in a fraction. It shows the number of equal parts.

Example:

$\frac{3}{4}$ ← numerator $\frac{3}{4} = \frac{1}{4} + \frac{1}{4} + \frac{1}{4}$

O

obtuse angle An angle greater than a right angle and less than a straight angle.

obtuse triangle A triangle with one obtuse angle.

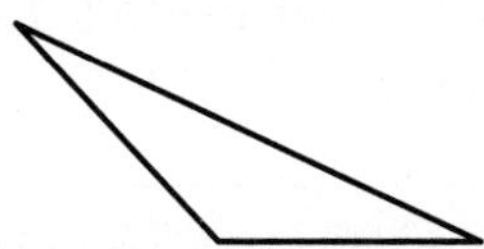

opposite sides Sides that are across from each other; they do not meet at a point.

Example: Sides *a* and *c* are opposite.

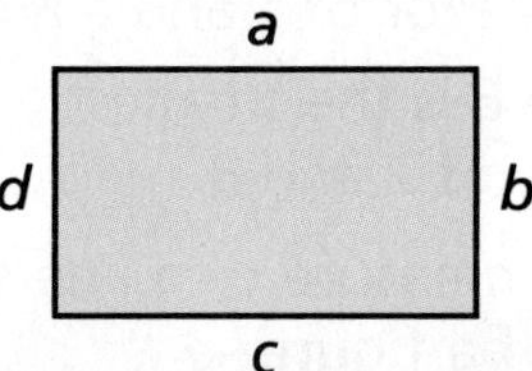

Order of Operations A set of rules that state the order in which operations should be done.

STEPS: -Compute inside parentheses first.
-Multiply and divide from left to right.
-Add and subtract from left to right.

ounce A unit of weight.
16 ounces = 1 pound
A unit of liquid volume (also called a fluid ounce).
8 ounces = 1 cup

P

parallel Lines in the same plane that never intersect are parallel. Line segments and rays that are part of parallel lines are also parallel.

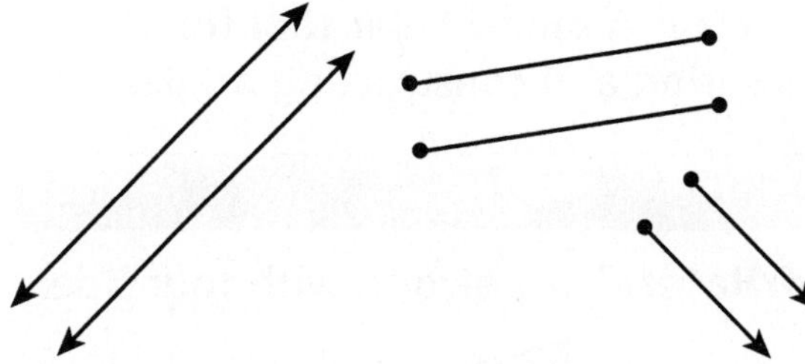

parallelogram A quadrilateral with both pairs of opposite sides parallel.

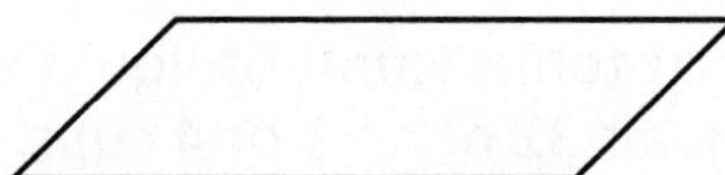

partial product The product of the ones, or tens, or hundreds, and so on in multidigit multiplication.

Example:

```
  24
×  9
----
  36  ←  partial product (9 × 4)
 180  ←  partial product (9 × 20)
----
 216
```

perimeter The distance around a figure.

perpendicular Lines, line segments, or rays are perpendicular if they form right angles.

Example: These two lines are perpendicular.

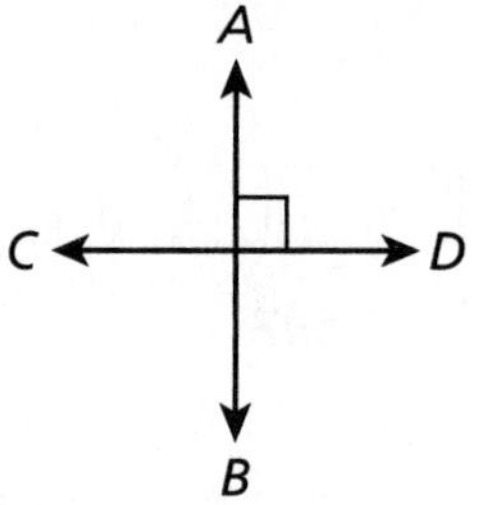

pictograph A graph that uses pictures or symbols to represent data.

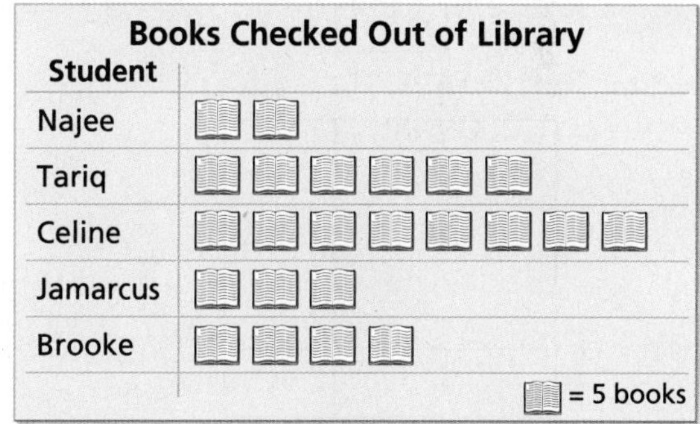

pint A customary unit of liquid volume that equals 16 fluid ounces.

place value The value assigned to the place that a digit occupies in a number.

Example: 235

The 2 is in the hundreds place, so its value is 200.

place value drawing A drawing that represents a number. Thousands are represented by vertical rectangles, hundreds are represented by squares, tens are represented by vertical lines, and ones by small circles.

Example:

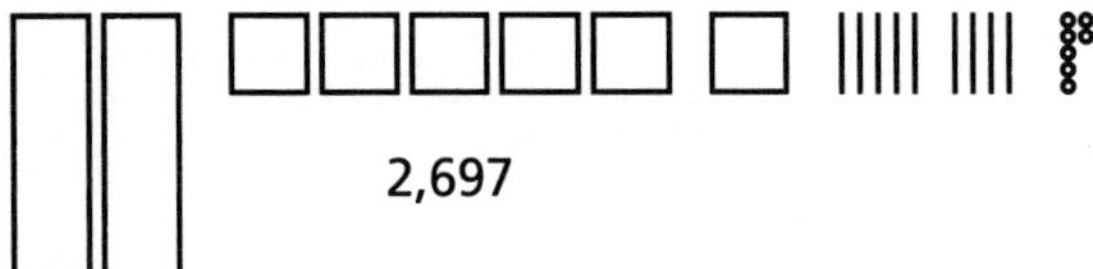

Place Value Sections A method using rectangle drawings to solve multiplication or division problems.

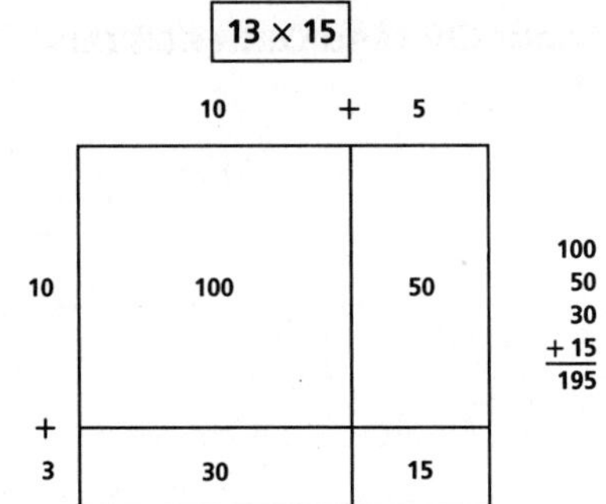

330 ÷ 5

a.
5 | 330

b.
60
5 | 330
− 300
30

c.
60 +
5 | 330
− 300
30

d.
60 +
5 | 330 | 30
− 300
30

e.
60 + 6
5 | 330 | 30
− 300 | − 30
30

f.
60 + 6 = 66
5 | 330 | 30
− 300 | − 30
30 | 0

point A location in a plane. It is usually shown by a dot.

polygon A closed plane figure with sides made of straight line segments.

pound A unit of weight in the U.S. customary system.

prefix A letter or group of letters placed before a word to make a new word.

prime number A number greater than 1 that has 1 and itself as the only factor pair. Examples of prime numbers are 2, 7, and 13. The only factor pair of 7 is 1 and 7.

product The answer to a multiplication problem.

Example: $9 \times 7 = 63$
↑ product

protractor A semicircular tool for measuring and constructing angles.

Q

quadrilateral A polygon with four sides.

quart A customary unit of liquid volume that equals 32 ounces or 4 cups.

quotient The answer to a division problem.

Example: $9\overline{)63}$ with 7 above; 7 is the quotient.

R

ray Part of a line that has one endpoint and extends without end in one direction.

rectangle A parallelogram with four right angles.

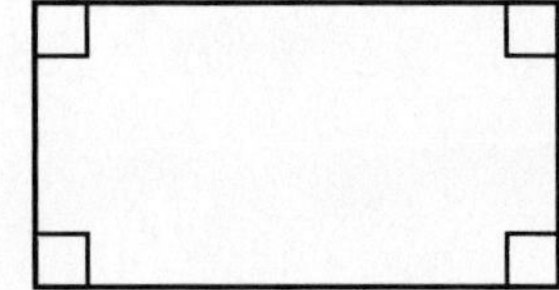

reflex angle An angle with a measure that is greater than 180° and less than 360°.

remainder The number left over after dividing two numbers that are not evenly divisible.

Example: $\begin{array}{r} 8 \text{ R3} \\ 5\overline{)43} \end{array}$ The remainder is 3.

rhombus A parallelogram with sides of equal length.

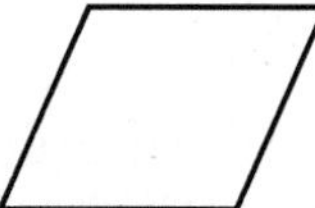

right angle One of four angles made by perpendicular line segments.

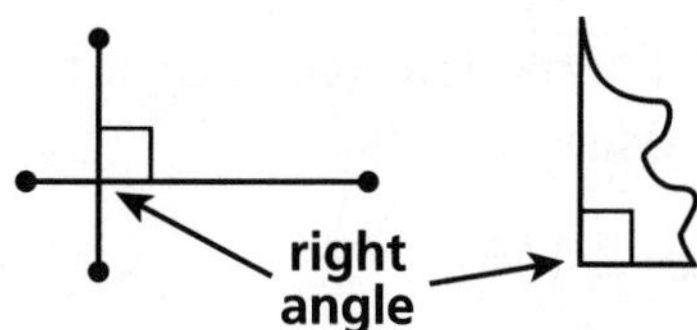

right triangle A triangle with one right angle.

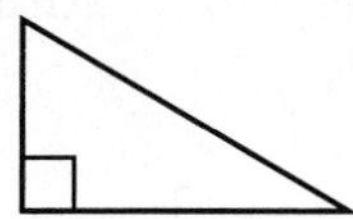

round To find the nearest ten, hundred, thousand, or some other place value. The usual rounding rule is to round up if the next digit to the right is 5 or more and round down if the next digit to the right is less than 5.

Examples: 463 rounded to the nearest ten is 460.
463 rounded to the nearest hundred is 500.

row A part of a table or array that contains items arranged horizontally.

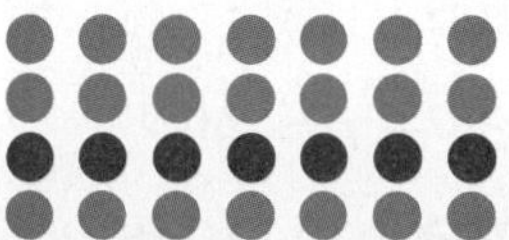

S

scalene A triangle with no equal sides is a scalene triangle.

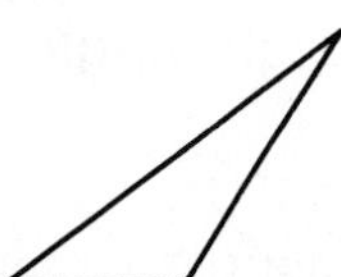

Shortcut Method A strategy for multiplying. It is the current common method in the United States.

Step 1	Step 2
$\begin{array}{r} {}^{7} \\ 28 \\ \times\ 9 \\ \hline 2 \end{array}$	$\begin{array}{r} {}^{7} \\ 28 \\ \times\ 9 \\ \hline 252 \end{array}$

simplest form A fraction is in simplest form if there is no whole number (other than 1) that divides evenly into the numerator and denominator.

Example: $\frac{3}{4}$ This fraction is in simplest form because no number divides evenly into 3 and 4.

simplify an expression Combine like terms and perform operations until all terms have been combined.

simplify a fraction To divide the numerator and the denominator of a fraction by the same number to make an equivalent fraction made from fewer but larger unit fractions.

Example: $\frac{5}{10} = \frac{5 \div 5}{10 \div 5} = \frac{1}{2}$

situation equation An equation that shows the structure of the information in a problem.

Example: $35 + n = 40$

solution equation An equation that shows the operation that can be used to solve the problem.

Example: $n = 40 - 35$

square A rectangle with 4 sides of equal length and 4 right angles. It is also a rhombus.

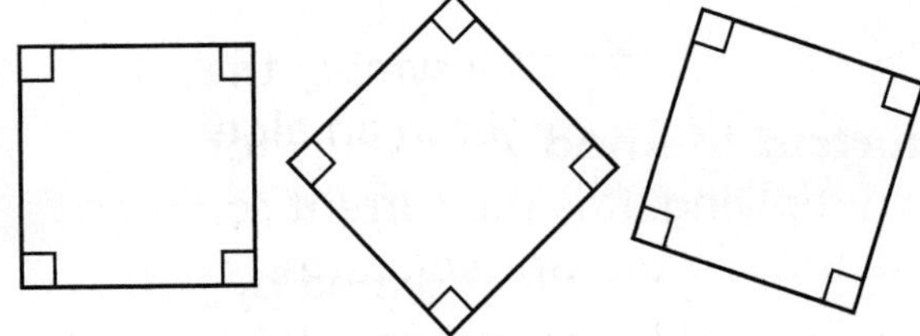

square array An array in which the number of rows equals the number of columns.

square centimeter A unit of area equal to the area of a square with one-centimeter sides.

square decimeter A unit of area equal to the area of a square with one-decimeter sides.

square foot A unit of area equal to the area of a square with one-foot sides.

square inch A unit of area equal to the area of a square with one-inch sides.

square kilometer A unit of area equal to the area of a square with one-kilometer sides.

square meter A unit of area equal to the area of a square with one-meter sides.

square mile A unit of area equal to the area of a square with one-mile sides.

square millimeter A unit of area equal to the area of a square with one-millimeter sides.

square unit A unit of area equal to the area of a square with one-unit sides.

square yard A unit of area equal to the area of a square with one-yard sides.

standard form The form of a number written using digits.

Example: 2,145

straight angle An angle that measures 180°.

sum The answer when adding two or more addends.

Example:

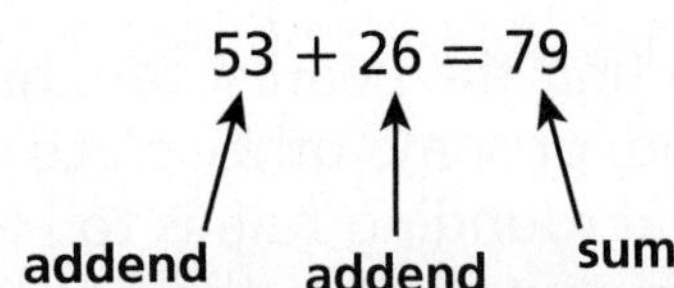

T

table Data arranged in rows and columns.

tenth A unit fraction representing one of ten equal parts of a whole, written as 0.1 or $\frac{1}{10}$.

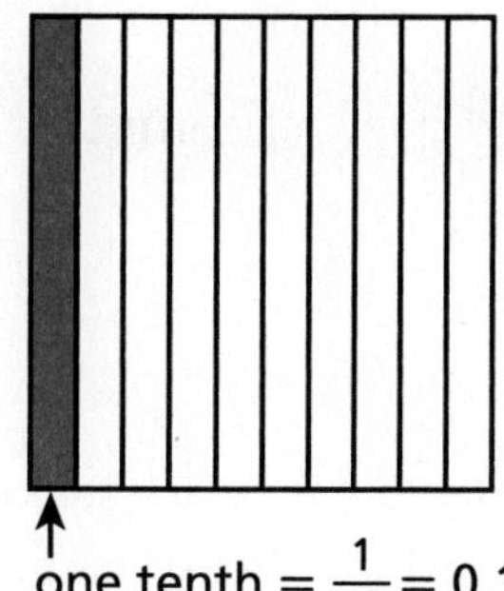

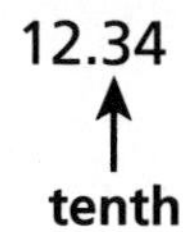

term in an expression A number, variable, product, or quotient in an expression. Each term is separated by an operation sign (+, −).

Example: $3n + 5$ has two terms, $3n$ and 5.

thousandth A unit fraction representing one of one thousand equal parts of a whole, written as 0.001 or $\frac{1}{1{,}000}$.

ton A unit of weight that equals 2,000 pounds.

tonne A metric unit of mass that equals 1,000 kilograms.

total Sum. The result of addition.

Example:

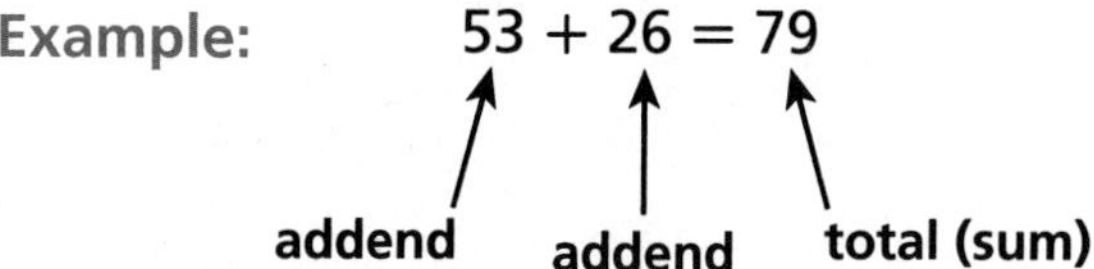

trapezoid A quadrilateral with exactly one pair of parallel sides.

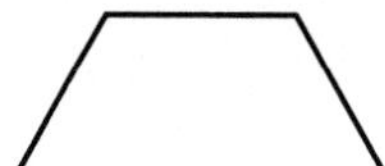

triangle A polygon with three sides.

U

unit A standard of measurement.

Examples: Centimeters, pounds, inches, and so on.

unit fraction A fraction whose numerator is 1. It shows one equal part of a whole.

Example: $\frac{1}{4}$ 

V

variable A letter or a symbol that represents a number in an algebraic expression.

vertex A point that is shared by two sides of an angle or two sides of a polygon.

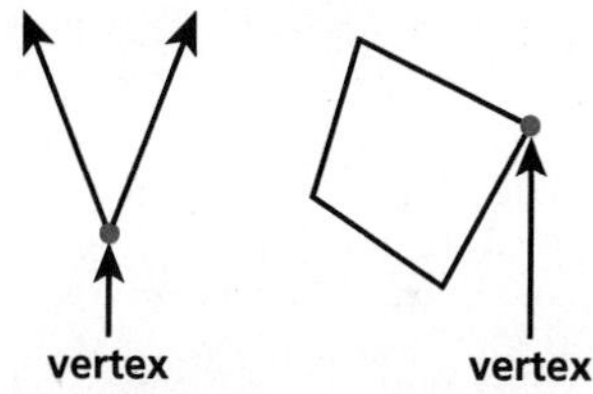

W

width The measure of one side of a figure.

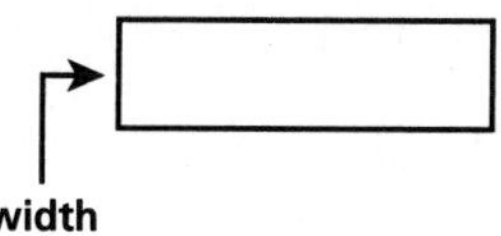

word form The form of a number written using words instead of digits.

Example: Six hundred thirty-nine

Y

yard A U.S. customary unit of length equal to 3 feet.